Agricultural Trade Policy Reforms and Trade Liberalisation in the Mediterranean Basin

Schriften zur Internationalen Entwicklungs- und Umweltforschung

Herausgegeben vom

Zentrum für internationale Entwicklungs- und Umweltforschung

der Justus-Liebig-Universität Gießen

Band 24

PETER LANG

Frankfurt am Main · Berlin · Bern · Bruxelles · New York · Oxford · Wien

Aikaterini Kavallari

Agricultural Trade Policy Reforms and Trade Liberalisation in the Mediterranean Basin

A Partial Equilibrium Analysis
of Regional Effects on the EU-27
and on the Mediterranean Partner Countries

PETER LANG
Internationaler Verlag der Wissenschaften

Bibliographic Information published by the Deutsche Nationalbibliothek
The Deutsche Nationalbibliothek lists this publication in the Deutsche Nationalbibliografie; detailed bibliographic data is available in the internet at <http://www.d-nb.de>.

Zugl.: Gießen, Univ., Diss., 2008

D 26
ISSN 1615-312X
ISBN 978-3-631-59203-8
© Peter Lang GmbH
Internationaler Verlag der Wissenschaften
Frankfurt am Main 2009
All rights reserved.

All parts of this publication are protected by copyright. Any utilisation outside the strict limits of the copyright law, without the permission of the publisher, is forbidden and liable to prosecution. This applies in particular to reproductions, translations, microfilming, and storage and processing in electronic retrieval systems.

www.peterlang.de

Preface

Trade integration around the Mediterranean is nowadays a particularly updated topic as it is directly related to the discussions of creating a Mediterranean Union, an initiative of the French President stemming out the Euro- Mediterranean Association Agreements, which aimed to form a Free Trade Area between the European Union and the non-EU Mediterranean Countries. The French initiative should be seen from the background of recent discussions focusing on the EU east enlargement and the relationships of the EU with its new eastern neighbours.

The southern EU Member States fear that the EU east enlargement and the opening of the markets with the new eastern neighbours will be particularly beneficial for the northern EU countries, as these Member States due to geographic proximity and historical relations will expand their trade towards east. Moreover they worry that resources from the EU budget will be transferred to the new Member States to support the agricultural markets and their re-structuring. Besides it has been often criticised that the northern agricultural products have received a higher protection compared to the Mediterranean ones, implying that the farmers in the northern of the Union have received higher income support and are better off than the ones in the southern of the EU. The French idea will certainly activate similar positive processes in the southern EU Member States.

The Mediterranean Partner Countries are on the one side interested in expanding their preferential access to the EU markets because this is promising for boosting their exports to the EU but on the other side fear of having their markets overwhelmed with imported commodities from the EU, for which still high custom duties apply. These contradictory effects create uncertainty regarding the future of the Mediterranean agriculture. How trade flows will evolve, who will benefit, who will be worse off, what the changes for the taxpayers will be depend from the magnitude of the forthcoming or recently agreed reforms and cannot be answered without an empirical analysis.

Based on an extended overview of the agricultural trade flows around the Mediterranean basin, Aikaterini Kavallari examines the effects of the recent reforms of the EU's Common Agricultural Policy, the EU enlargement and the Euro-Mediterranean Agreements on the agricultural sector of the Mediterranean countries. The simulations are run with an extended and adjusted version of the partial equilibrium trade policy model AGRISIM. Objective is to quantify the allocative and distributional effects of these reforms on the EU Member States and its Mediterranean Partner Countries and to conclude on the competitiveness of the region.

Overall the study can be considered as very successful since it penetrates not only theoretically but mainly empirically with the own simulations a highly complex theme, this of trade integration in the Mediterranean. It is certainly a contribution that touches a politically interesting topic, since the negotiations within the Euro-Mediterranean Association Agreements are still running and therefore is particularly relevant and telling for the policy makers.

Giessen, January 2009 P. Michael Schmitz

Acknowledgments

My PhD thesis has been accomplished during the time I worked as research assistant at the Institute of Agricultural Policy and Market Research, Chair of Agricultural Policy and Development Economics of the Justus Liebig University of Giessen. To all the people, who helped me and contributed to this work I would like to express my deep gratitude and appreciation.

First of all I would like to express my gratitude to my first supervisor, Prof. Dr. P. Michael Schmitz, whose contribution to every step of this work is inestimable. He initiated me to the world of modelling and made for me possible to accomplish my dissertation. His clear view, constructive criticism and his constant readiness to give advice and guidance to all problems that arose have encouraged me to carry on, to complete this work and have contributed to have a pleasant time in Giessen even beyond the work of my thesis.

Equally I would like to thank Prof. Dr. Roland Herrmann for being the second supervisor of my thesis. His constructive suggestions on my work and his support during the last stages of my work have been very helpful.

I am also indebted to my academic teacher Prof. Dr. Konstantinos Mattas, who is standing by me and is supporting me since the early stages of my academic life as undergraduate student of the Aristotelian University of Thessaloniki. I own him gratitude for enabling me to participate in the EU research project MEDFROL, from which I received financial support to develop the empirical part of my thesis. His insightful and detailed criticism on early stages of my work is highly appreciated. I would also like to thank the consortium of the MEDFROL project and especially Prof. Dr. Yves Surry for being always very enthusiastic and ready to offer me useful comments.

I own debt to all my colleagues in the Institute of Agricultural Policy and Market Research of the University of Giessen and particularly to Joachim Hesse, Sirirat Kiatpathomchai, Nataliya Möser, Juliane Stoll, Janine Wronka, as well as to my ex-colleagues Kim Schmitz and Tobias Wronka for their technical and instrumental assistance. My dearest thanks are to René Borresch, not only for supporting me in programming AGRISIM in GAMS, for helping me with small and large last minute problems and for providing me with thorough comments on my work but also for tolerating me while we shared the same office for about a year and for being the best room mate I could ever wish. There are not enough words to thank Margot Hilla, who was for me much more than a helpful colleague. She made me part of her family, made her friends my friends, gave me all her love and attention and treated me as her own daughter.

Thanks are due to Dr. Nick Parrott for the grammatical editing of my thesis as well as to my colleagues in the Institute for Prospective Technological Studies of the European Commission's Joint Research Centre Dr. Thomas Fellmann, Dr. S. Hubertus

Gay, Dr. Ignacio Perez Dominguez and Dr. Alexander Stein for supporting me in my several ad-hoc lay-out enquiries while preparing the manuscript for the publication. Remaining errors are, of course, mine.

I would like to express my thanks to my dear friends Penelope Anagnostopoulou, Panagiotis Bargiotas, Vicky Pappa, as well as to my godson Efthimios Bargiotas for encouraging me, for supporting me during this period and for being always here the time I needed them. I could not ask for better friends.

Lastly and most importantly I would like to thank my family. I owe a debt of gratitude to my parents Charikleia and Panagiotis for their love and support not only throughout the course of my PhD studies but throughout my entire life. Without their encouragement I would not have pursued postgraduate studies. My brothers Dimitrios and Paraskevas as well as my sisters-in-law Marina and Myrto offered me moral support and stood by me in times of difficult decisions. This book is dedicated to my parents.

Giessen, January 2009																																Aikaterini Kavallari

Table of contents

LIST OF ABBREVIATIONS	XI
LIST OF TABLES	XV
LIST OF FIGURES	XIX
1	**INTRODUCTION** ... 1
1.1	ISSUS AND OBJECTIVES OF THE STUDY ... 1
1.2	STRUCTURE OF THE STUDY ... 3
2	**AGRICULTURAL TRADE IN THE MEDITERRANEAN BASIN** ... 5
2.1	THE DEVELOPMENT OF AGRICULTURAL TRADE IN THE MEDITERRANEAN BASIN ... 5
2.2	TRADE AGREEMENTS ... 12
2.3	DEVELOPMENT OF TRADE PROTECTION AMONG THE MEDITERRANEAN COUNTRIES ... 19
2.4	REFORMS OF AGRICULTURAL POLICIES THAT AFFECT TRADE ... 28
3	**THEORETICAL ANALYSIS OF AGRICULTURAL TRADE POLICIES** ... 33
3.1	CAP REFORM: ALLOCATIVE AND DISTRIBUTIONAL EFFECTS ... 34
3.2	FREE TRADE AREAS: ALLOCATIVE AND DISTRIBUTIONAL EFFECTS ... 37
3.3	MULTILATERAL LIBERALISATION: ALLOCATIVE AND DISTRIBUTIONAL EFFECTS IN A SINGLE MARKET ... 46
3.4	MARKET LINKAGES ... 53
4	**EMPIRICAL STUDIES OF AGRICULTURAL TRADE POLICY REFORMS IN THE MEDITERRANEAN BASIN** ... 61
4.1	BASIC ELEMENTS OF MODELLING TOOLS ... 61
4.2	LITERATURE REVIEW ... 65
4.2.1	*Empirical assessments of the latest CAP Reform relevant to Mediterranean countries* ... 65
4.2.2	*Empirical assessments of trade liberalisation that is relevant to Mediterranean countries* ... 67
4.2.3	*Outlook on empirical studies and identification of further research needs* ... 91
4.3	OVERVIEW OF AGRISIM ... 93
4.3.1	*Model description* ... 93

4.3.2	Calibration of Elasticities	104
4.3.3	Database	105
4.3.4	Technical Issues	107
4.3.5	Limitations of the model	108
5	**SIMULATIONS WITH AGRISIM AND MODEL RESULTS**	**111**
5.1	OVERVIEW OF THE BASELINE AND SIMULATION SCENARIOS	112
5.2	EFFECTS OF THE CAP REFORM FOR MEDITERRANEAN COMMODITIES AND THE ACCESSION OF BULGARIA AND ROMANIA	114
5.2.1	Effects on EU Member States	114
5.2.2	Effects on the MPCs	125
5.3	THE EFFECTS OF THE „BARCELONA AGREEMENT"	128
5.3.1	Effects on the EU Member States	128
5.3.2	Effects on the MPCs	133
5.4	THE EFFECTS OF MULTILATERAL LIBERALISATION	139
5.4.1	Effects on the EU Member States	139
5.4.2	Effects on the MPCs	149
5.5	THE COMPETITIVENESS OF MEDITERRANEAN COUNTRIES	155
6	**CONCLUDING REMARKS**	**161**
7	**SUMMARY**	**167**
LIST OF REFERENCES		**177**
ANNEX		**193**
A.	TO CHAPTER TWO	193
B.	TO CHAPTER FIVE	207

List of Abbreviations

AGRISIM	Agricultural Simulation Model
AMAD	Agricultural Market Access Database
AMU	Arab Maghreb Union
AVE	Ad-Valorem Equivalent
CAP	Common Agricultural Policy
CEEC-ASIM	Central and Easter European Countries – Agricultural Simulation Model
CEECs	Central and Eastern European Countries
CMO	Common Markets Organisation
CGE	Computable General Equilibrium model
COMESA	Common Market for Eastern and Southern Africa
COMTRADE	Commodity Trade Statistics Database of the United Nations
CSE	Consumer Support Estimate
DRC	Domestic Resource Cost
EAGGF	European Guidance and Guarantee Fund
ECO	Economic Cooperation Organisation
EFTA	European Free Trade Association countries
EU	European Union
FAOSTAT	Food and Agriculture Organisation of the United Nations Statistics
FAPRI	Food and Agricultural Policy Research Institute
FTA	Free Trade Area
GAFTA	Greater Arab Free Trade Area
GDP	Gross Domestic Product
GNI	Gross National Input
GSP	Generalised System of Preferences

GSTP	General System of Trade Preferences
GTAP	Global Trade Analysis Project
IAMO	Leibniz Institute of Agricultural Development in Central and Eastern Europe
IO	Input Output
MENA	Middle East and North African countries
MFN	Most Favourite Nation
MPCs	Mediterranean Partner Countries
NAFTA	North Atlantic Free Trade Area
NGQ	National Guaranteed Quantity
NPR	Nominal Protection Rate
NTB	Non Tariff Barrier
NUTS	Nomenclature des Unités Territoriales Statistiques
OECD	Organisation for the Economic Cooperation and Development
PSE	Producer Support Estimate
PE	Partial Equilbrium
RTA	Regional Trade Agreement
SAM	Social Accounting Matrix
SWOPSIM	Static World Policy Simulation Modelling Framework
t	tonnes
TARIC	Tarif Intégré des Communautés Européennes (integrated tariff database of the European Union)
TRAINS	Trade Analysis and Information System
UNCTAD	United Nations Conference on Trade and Development
USDA	United States Department of Agriculture
VAT	Value Added Tax

List of abbreviations

VPM   Value of Preference Margin

WTO   World Trade Organisation

List of Tables

Table 2.1: Share of MPCs in the extra EU-25 trade of agricultural commodities (HS01-24) excluding fish (HS03) 10

Table 2.2: Status of the Association Agreements between the EU and the MPCs as of January 2008 14

Table 2.3: WTO status, bilateral and regional agreements of the MPCs as in January 2008 17

Table 2.4: Weighted averages of ad-valorem equivalents of applied tariffs by the EU from imports from the MPCs in 2005 20

Table 2.5: Weighted averages of ad-valorem equivalents of applied tariffs by the MPCs to imports from the EU in 2005 25

Table 4.1: Overview of ex-ante empirical studies on modelling agricultural trade policy liberalisation on the Mediterranean Basin with equilibrium models 71

Table 4.2: Commodities' and countries' list 105

Table 5.1: Levels of aggregation used in the AGRISIM Database 111

Table 5.2: Base run and simulated scenarios 113

Table 5.3: Net Protection Rate in the EU-25 markets, in % 116

Table 5.4: Budgetary effects from changes in direct subsidies within the EU-25 (deviations from Base Run in US$ million) 118

Table 5.5: Budgetary effects of changes in the decoupling of direct subsidies in the EU-27 (all agricultural markets, deviations from Base Run in US$ million) 118

Table 5.6: Budgetary effects of changes in customs duties in the EU-25 (deviations from Base Run in US$ million) 119

Table 5.7: Budgetary effects of changes in direct subsidies and customs duties in Bulgaria and Romania (deviations from Base Run in US$ million) ...120

Table 5.8: Budgetary changes in the EU-25 due to the CAP Reform for Mediterranean commodities (SC1) and the accession of Bulgaria and Romania (SC2) (in US$ million) 121

Table 5.9: Net protection rates in the MPCs, in % 126

Table 5.10: Changes in the customs duties of the MPCs due to the CAP reform of Mediterranean commodities (SC1) and the accession of Bulgaria and Romania to the EU (SC2) (deviations from Base Run in US$ million) 127

Table 5.11: Net protection rates in the EU-25 markets, in % 131

Table 5.12: Change in customs revenues in the EU-27 due to the „Barcelona Agreement" (SC3) (deviations from new Base Run (SC2) in US$ million) 132

Table 5.13: Welfare effects in the EU-27 due to the „Barcelona Agreement" (SC3) (deviations from new Base Run (SC2) in US$ million) 133

Table 5.14: Net protection rates in MPCs, in % 136

Table 5.15: Changes of customs duties in MPCs due to the „Barcelona Agreement" (SC3) (deviations from new Base Run (SC2) in US$ million) 138

Table 5.16: Welfare effects in MPCs due to the „Barcelona Agreement" (SC3) (deviations from new Base Run (SC2) in US$ million) 139

Table 5.17: Net protection rates in the EU-25 markets, in % 143

Table 5.18: Changes in direct subsidies in the EU-27 due to „WTO liberalisation" (SC4) (deviations from new Base Run (SC2) in US$ million) 144

Table 5.19: Changes in direct subsidies in the EU-27 due to full liberalisation (SC5) (deviations from new Base Run (SC2) in US$ million) 145

Table 5.20: Changes in input subsidies in the EU-27 due to „WTO liberalisation" (SC4) (deviations from new Base Run (SC2) in US$ million) 145

Table 5.21: Changes in input subsidies in the EU-27 due to full multilateral trade liberalisation (SC5) (deviations from new Base Run (SC2) in US$ million) 146

Table 5.22: Changes in customs duties in the EU-27 due to „WTO liberalisation" (SC4) (deviations from new Base Run (SC2) in US$ million) 147

Table 5.23: Changes in customs duties in the EU-27 due to full multilateral trade liberalisation (SC5) (deviations from new Base Run (SC2) in US$ million) 147

Table 5.24: Welfare effects in the EU-27 due to „WTO liberalisation" (SC4) (deviations from new Base (SC2) in US$ million) 148

Table 5.25: Welfare effects on the EU-27 due to full multilateral trade liberalisation (SC5) (deviations from new Base Run (SC2) in US$ million) 148

List of tables XVII

Table 5.26: Net protection rates in the MPCs' markets, in % 152

Table 5.27: Changes in customs duties in MPCs due to „WTO liberalisation" (SC4) and full multilateral trade liberalisation (SC5) (deviations from new Base Run (SC2) in US$ million) 153

Table 5.28: Welfare effects on MPCs due to „WTO liberalisation" (SC4) and full multilateral trade liberalisation (SC5) (deviations from SC2 in US$ million) 154

Table 7.1: Overview of simulated scenarios 171

Table 7.2: Evaluation of policy scenarios on EU Mediterranean Member States .. 174

Table 7.3: Evaluation of policy scenarios on northern EU Member States 174

Table 7.4: Evaluation of policy scenarios on new EU Member States (EU-10) 175

Table 7.5: Evaluation of policy scenarios on the MPCs 175

Table A.1: Value of Preference Margins resulting from the Euro-Mediterranean Association Agreements in '000 US$ (1999) 196

Table A.2: Value of Preference Margins resulting from the Euro-Mediterranean Association Agreements in '000 US$ (2003) 197

Table A.3: Applied tariffs by the MPCs (in %) 198

Table A.4: Export subsidies for agricultural commodities reported by the MPCs to the WTO 202

Table B.1: Base Run projections of product balances and prices 207

Table B.2: Custom duties and agricultural subsidies in Base Year (in US$ million) 210

Table B.3: Changes in commodity balances (deviation from BA in %) 214

Table B.4: Net trade effects (in 1000 t) 218

Table B.5: Changes in prices (deviation from BA in %) 222

Table B.6: Changes of farmer's revenue (% deviations from BA) 228

Table B.7: Changes in production shares relative to the world supply (in %) 231

Table B.8: Change in production shares of the Mediterranean Member States relative to the EU-27 supply (in %) 233

Table B.9: Change in welfare (deviations from BA in US$ million) 234

List of figures

Figure 2.1: Main trading partners of the EU-25, as a % of external EU trade in 2005 6

Figure 2.2: Main trade partners of MPCs, as % of their total trade in 2004 7

Figure 2.3: Share of the EU-25 in the external trade of MPCs in %, 2000 and 2004 8

Figure 2.4: Value of Preference Margin for selected agricultural commodities resulting from the Euro-Mediterranean Association Agreements 23

Figure 2.5: Nominal protection rate in %, average of 2000-2002 27

Figure 3.1: Impacts of different levels of production-effectiveness of direct payments on producer's welfare 36

Figure 3.2: Effects of countries A and B forming an FTA, where countries B and C have the same production costs and the post-FTA demand of country A is met by imports from both B and C 41

Figure 3.3: Effects of countries A and B forming an FTA, where countries B and C have the same production costs and the post-FTA demand of country A is met by imports from B alone 42

Figure 3.4: Welfare effects of trade liberalisation on a net importing country 47

Figure 3.5: Welfare effects of trade liberalisation on a net exporting country 48

Figure 3.6: Welfare effects of agricultural trade liberalisation on small countries .. 49

Figure 3.7: Preference erosion effects of multilateral agricultural liberalisation in the two partner countries A and B and in the third country C 52

Figure 3.8: Vertical linkages between two agricultural markets 54

Figure 3.9: Horizontal linkages between two agricultural markets 58

Figure 4.1: Simulations-routine in AGRISIM; example of 2 markets – 2 commodities 94

Figure 4.2: The modules of AGRISIM and their inter-connections 107

Figure 5.1: Net trade effects of the CAP Reform for Mediterranean commodities on Spain, Greece and Italy 115

Figure 5.2: Allocation of welfare effects among the EU Mediterranean Member States due to the CAP Reform for Mediterranean commodities (SC1) and the accession of Bulgaria and Romania (SC2) (deviations from BA) 122

Figure 5.3: Allocation of welfare effects on the rest of the EU-15 and 2004 accession states as a result of the CAP Reform of Mediterranean

		commodities (SC1) and the accession of Bulgaria and Romania (SC2) (deviations from BA) 123
Figure 5.4:		Welfare effects on Bulgaria and Romania due to the CAP Reform for the Mediterranean commodities (SC1) and their accession to the EU (SC2) (deviations from BA) 124
Figure 5.5:		Net trade effects of the „Barcelona Agreement" on Spain, Greece and Italy (SC3) 129
Figure 5.6:		Net trade effects of the „Barcelona Agreement" on non-Mediterranean EU regions (SC3)... 130
Figure 5.7:		Net trade effects of the „Barcelona Agreement" on MPCs (SC3)....... 135
Figure 5.8:		Net trade effects of „WTO liberalisation" on Mediterranean EU Member States (SC4) 141
Figure 5.9:		Net trade effects of „WTO liberalisation" on non-Mediterranean EU regions (SC4).. 142
Figure 5.10:		Net trade effects on MPCs due to „WTO liberalisation" (SC4) 150
Figure 5.11:		Net trade effects of the „Barcelona Agreement" (SC3) and „WTO liberalisation" on Spain, Greece and Italy (SC4) 156
Figure 5.12:		Net trade effects of the „Barcelona Agreement" (SC3) and „WTO liberalisation" (SC4) on Morocco, Turkey and the other MPCs 157
Figure 5.13:		Changes in the world supply shares of EU Mediterranean Member States (in %)... 158
Figure 5.14:		Changes in world supply shares of all MPCs (in %)......................... 159
Figure 5.15:		Changes in world supply shares of typical Mediterranean commodities and of apples (deviations from new Base Run (SC2) in percentage points).. 160
Figure A.1:		Illustration of preference margin effects 194

1 Introduction

1.1 Issus and objectives of the study

Bilateral and multilateral trade agreements have been gaining attention and relevance in recent years. After the Doha negotiations came to a deadlock in the summer of 2006 international attention has been increasingly focused on regional trade integration. The initiatives of the EU to conclude or to deepen existing trade agreements with blocks of countries in Africa, Asia and the Pacific have been a particular focus of attention.

There are a number of regional trade agreements within the Mediterranean basin that are the result of traditional trade relationships and which have led to deeper trade integration. Of these, the most significant is the Barcelona Agreement, an initiative of the EU which started in 1995 and aims to establish a Free Trade Area (FTA) with ten non-EU countries known as the Mediterranean Partner Countries (MPCs) by 2010[1].

This agreement is of special relevance to EU agriculture, because the MPCs compete directly with the EU's Mediterranean Member States for market share in northern EU countries (AQUILA and VELAZQUEZ, 2002). This is not the case for the other regional trade agreements in which the EU is involved, such as the Economic Partnership Agreement (EPAs) between the EU and the African, Caribbean and Pacific (ACP) group of countries (Cotonou Agreement of 2000), the European Free Trade Association (EFTA) countries, or the negotiations with the Southern Cone Common Market (MERCOSUR) countries. This creates a divergence of interests between the northern and the Mediterranean EU Member States over further liberalising trade between the EU and the MPCs. The northern EU Member States expect to benefit from improved market access to North African and Middle East countries for "typical northern" commodities, such as cereals, dairy products and meat and thus support the Euro-Med Agreements. By contrast the Mediterranean EU Member States fear losing their market shares in the fruit and vegetables markets in northern EU countries and thus oppose the liberalisation of trade in agricultural commodities trade around the Mediterranean. The MPCs specialise in producing fruits and vegetables, mainly citrus fruits, dates and tomatoes, in which they have comparative advantages (GALANOPOULOS et al., 2007), and mainly import cereals and livestock commodities.

On the other side of the Mediterranean Sea, the MPCs expect to benefit from the Euro-Med Agreements through gaining better market access for their export com-

---

[1] The Barcelona Agreement was signed in 1995 at the EU Summit of Barcelona by the (then) 15 Member Countries of the EU and 12 Mediterranean Countries Algeria, Cyprus, Egypt, Jordan, Israel, Lebanon, Malta, Morocco, Syria, the Gaza Strip and the West Bank, Tunisia and Turkey. These countries apart from Cyprus and Malta (which have subsequently joined the EU) are hereafter called Mediterranean Partner Countries (MPCs).

modities, but in parallel with this they will have to abolish tariffs on imported commodities and thus lose the import rent. Due to low competitiveness with "northern" products they also fear that their markets will be overwhelmed with imported commodities. These contradictory effects lead the MPCs to see the forthcoming trade liberalisation with more scepticism than enthusiasm.

Parallel to the developments at the bilateral level there are ongoing discussions within the World Trade Organisation (WTO) on multilateral trade liberalisation. Although opening up markets is seen positively by economists as it enhances welfare, when it comes to preferential partners it is not clear who will be the winners and losers. This is because multilateral liberalisation is associated with preference erosion effects and it is not clear whether trade creation will be greater than these trade diversion effects. This question is of relevance for both the EU and the MPCs and empirical evidence is needed to inform the negotiations over future liberalisation of trade in agricultural commodities.

Finally the agricultural sector around the Mediterranean is highly influenced by the recent reforms of the EU's Common Agricultural Policy. This directly affects those countries around the Mediterranean basin that are EU Member States, but also has an effect on the MPCs since the EU is their most important trade partner. Changes in production patterns within the EU are reflected in trade and will certainly have impacts on welfare and the allocation and distribution of resources in the MPCs as well.

Forthcoming and recently applied policy reforms are expected to change the level and structure of trade flows of agricultural commodities around the Mediterranean, affecting the competitiveness of the countries involved. This is expected to increase regional disparities not only among countries on the two coastal sides of the Mediterranean basin (the north and the south and east) but also within the EU, due to the different agricultural specialisations of northern and southern EU Member States. The Barcelona Agreement is expected to benefit the northern EU Member States because of low competition with the MPCs. Its effects upon EU and non-EU Mediterranean countries are expected to be more ambiguous, due to the high level of competition among them. Farmers within the EU's Mediterranean Member States fear a reduction of their production levels, of farm gate prices and thus of their income, and accuse the EU's agricultural policy of giving more support to northern EU farmers. These fears are in addition to their worries about how they will be able to cope with the new balance brought about by the recent reform of the EU's Common Agricultural Policy. Multilateral liberalisation will certainly benefit third countries but it is not clear whether the benefits from the Barcelona Agreement coming into full force will be equally shared, or indeed if some countries (both EU and MCP) will be net winners or losers.

Within this framework, the objective of this study is to give an overview of the existing status quo of agricultural trade around the Mediterranean and to provide deeper insights into the likely impacts of agricultural policy reforms in the Mediterranean basin. Two further objectives are to evaluate the likely effects of the new policy regimes (and suggest further changes) and to provide sound empirical results that can inform policy makers involved in discussing the future of Mediterranean agriculture. In particular the study attempts to address the following questions:

1. What are the likely overall impacts of trade liberalisation for the agricultural sector in the EU and in the MPCs?
2. What are the likely regional effects within the EU? Will the southern EU Member States be discriminated against in comparison with the northern ones by the changed agricultural policy?
3. What are the likely effects on producers and consumers in the two main trading blocks (i.e. the EU and MPCs)?
4. What are the likely budgetary effects for the EU as a whole and for particular Member States?

1.2   Structure of the study

The study can be characterised as an *ex-ante* analysis that follows the basic structure of a typical "what if" analysis. After an overview of the existing situation, follows a theoretical assessment of the effects of the forthcoming and recently changed agricultural policy reforms. This theory is then tested by applying an empirical model, the results of which are intended to be of use to policy makers.

Hence, the second chapter contains an overview of agricultural trade around the Mediterranean. It presents a detailed analysis of the development of trade flows, the most significant trade partners and commodities traded. This chapter also gives an overview of existing trade agreements among Mediterranean countries, which are a key determinate of the development of trade flows and affect the levels of trade protection applied and countries' trade preferences. For this purpose the value of preference margin and the nominal protection rate are calculated, which are applied for the most commonly traded agricultural commodities and indicate the trade preferences granted within the Barcelona Agreement and the trade protection respectively. Existing and proposed reforms of agricultural trade policies are presented in detail as these are the other feature influencing how agricultural trade around the Mediterranean is organised, as well as its likely future changes and challenges.

The effects of these agricultural policy reforms are theoretically assessed in the third chapter, using the theory of applied welfare economics. The allocational and distributional effects of the EU's Common Agricultural Policy, the full application of the Barcelona Agreement and multilateral trade liberalisation are shown for single markets using a static partial framework. Market linkages and their effects are discussed in the last section of this chapter.

A literature review of existing empirical assessments of agricultural policy reforms with relevance for Mediterranean countries follows in the fourth chapter. This chapter focuses on the methodologies applied, the scenarios examined and the results obtained. This review is used to identify gaps in knowledge, future research needs and to choose the most appropriate method for addressing the objectives of this study. Based on these findings the last section of this chapter discusses the AGRISIM model, selected as the most appropriate empirical research tool. This is a partial equilibrium model, synthetic and deterministic in nature with iso-elastic demand and supply functions. The model can be used to simulate the effects on production, consumption, trade, domestic prices, border prices, state's budget, consumer surplus, producer surplus and overall welfare. The extensions and updates of the model version used in this study were undertaken within the EU-project MEDFROL[2].

The simulations and the results from the model AGRISIM are presented in the fifth chapter. The simulated scenarios are based upon:

a) the latest reform of the EU's Common Agricultural Policy and the enlargement of the EU;

b) trade preferences within the Barcelona Agreement and

c) multilateral trade liberalisation and preference erosion effects

The simulation results are divided into three main sections, based on these three scenarios. The effects are presented separately for the EU regions and the MPCs. The last section of this chapter explores the effects of the changes under these different scenarios on the competitiveness of Mediterranean countries. Especially for the EU Member States, the budget effects are captured with a new module that was programmed in AGRISIM, which allowed to take into account the intra-community financial flows between the budget of the entire EU and of each Member State.

Concluding remarks are given in the sixth chapter, and a summary of the overall study follows in the last chapter.

---

[2] Financial support is gratefully acknowledged from the Project MEDFROL "Market and Trade Policies for the Mediterranean Agriculture: The case of fruit/vegetables and olive oil", SSPE-CT-2004-502459 (STREP), 6th Framework Programme of the EU, for the development of this version of AGRISIM.

2 Agricultural trade in the Mediterranean basin

Trade relationships among the countries surrounding the Mediterranean have existed since ancient years. This chapter provides an overview of the trade in agricultural products and the way it has evolved in recent years. The chapter continues with a review of trade agreements among Mediterranean countries, the measures of trade protection that have been applied and a detailed description of the reforms of agricultural trade policy, factors that both determine and are determined by existing trade flows.

2.1 The development of agricultural trade in the Mediterranean basin

Around the Mediterranean basin, two groups of countries can be clearly distinguished. On the northern side of the Mediterranean Sea there are the Member States of the EU, while on the southern and eastern sides are the ten Mediterranean Partner Countries that are connected to the EU through Association Agreements.

The MPCs have much stronger trade flows with the EU than between themselves (as shown clearly in Figures 2.1 and 2.2). This is the result of geographic proximity and of historical trade relationships between the countries surrounding the Mediterranean basin.

According to EUROSTAT data, the MPCs are important trading partners for the EU, with the volume of trade being equivalent to the rapidly developing Asian countries although less important than the EU's trade with the USA and EFTA countries (BOUZERGAN, 2007). In 2005 they provided about 8 % of commodities imported by the EU, and were the EU's fifth most important supplier, while taking about 9.5 % of extra EU exports, and being the third most important importer after the USA and EFTA, which have shares of 23.5 and 11.2 % of extra-EU exports respectively. The same report states that MPCs' share of the EU's external exports remained quite stable over the period 2000-2005, whereas its share of the EU's external imports increased by one percent over the same period.

The EU has a positive trade balance with the MPCs as a group (approximately 13 billion Euro in 2005) and with all the individual MPC countries, apart from Algeria and Syria. Nevertheless the trade surplus fell by about 5 billion Euro between 2000 and 2005 because imports to the EU rose quicker than exports from the EU (by +37.4 % and +22.9 % respectively), as EUROSTAT data show. This is attributed to the development of the Euro-Med Agreements, one of the main aims of which was to promote imports into the EU. Indeed, as BOUZERGAN (2007) reported, during 2000-2005 the annual growth rate of imports from MPCs into the EU was 6.5 %, while exports to the MPCs rose by 4.2 % p.a.. Overall MPCs share of the EU's external trade rose by one percent over the period 2000-2005 (EUROSTAT, 2006).

Within the EU the main trading partners of the MPCs are France, Germany and Italy. Trade links between the MPCs and the ten new Member States of the (former) EU-25 are quite limited but the trade volumes with certain countries, such as Hungary, Poland and the Czech Republic have shown a promising growth after their entry into the EU (BOUZERGAN, 2007).

Figure 2.1: Main trading partners of the EU-25, as a % of external EU trade in 2005

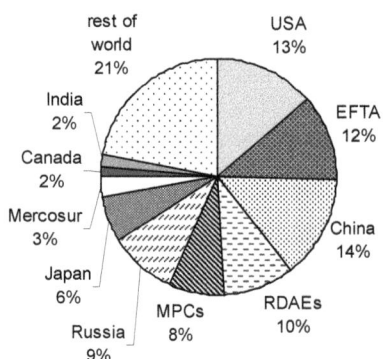
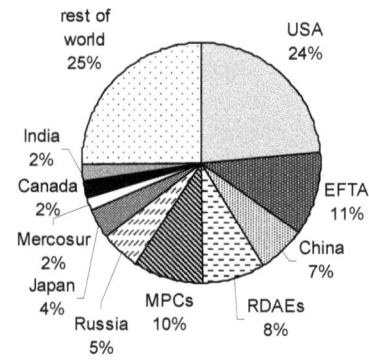

Total imports' value: 1,184 billions of Euro   Total exports' value: 1,072 billions of Euro

Notes: EFTA: Iceland, Liechtenstein, Norway, Switzerland; Mercosur (South American Common Market): Argentina, Brazil, Paraguay and Uruguay; RDAEs (Rapidly Developing Asian Economies): Hong Kong, South Korea, Malaysia, Singapore, Thailand and Taiwan.
Source: Own compilation based on BOUZERGAN, 2007

Conversely the EU is the most important trade partner for the MPCs, with about half of their exports being destined to EU markets and more than half of their imports coming from EU countries (BOUZERGAN, 2007; EUROSTAT, 2006). In 2004 products from the EU accounted for about 45.1 % of all imports made by the MPCs and about 48.7 % of the MPCs' exports were sold in EU markets. Other important trade partners for the MPCs are the USA and other Asian countries but these are much less important in terms of the MPCs' imports (approximately 7 % and 15 % respectively) and their exports (about 17 % and 11 % respectively), as shown in Figure 2.2.

Agricultural trade in the Mediterranean basin

Consequently, better market access to EU markets appears very attractive to the MPCs since it could give them the chance to expand their trade and to benefit from the effects of opening their economies.

It is worth noting that the volume of trade between the MPCs themselves is relatively modest. For example in 2004, intra-MPC trade accounted only for about 4.5 % of total imports into MPCs and nearly 6.2 % of their total exports. This could hinder further south-south integration which is required for the development of the Euro-Mediterranean Free Trade Area.

Figure 2.2: Main trade partners of MPCs, as % of their total trade in 2004

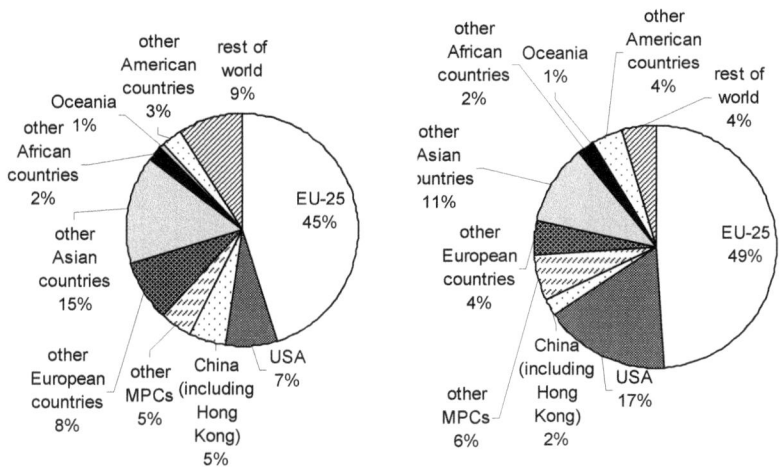

Total imports' value: 183.1 billion Euro   Total exports' value: 138.8 billion Euro

Source: Own compilation based on BOUZERGAN, 2007

There are notable differences in the MPCs' trading relationships with the EU (Figure 2.3). North African countries have a high proportion of their trade with the EU, but this share is much lower for Middle Eastern countries. For example, in 2004 Morocco and Tunisia sent about 83 % and 74 % of their exports to the EU respectively while just 3.2 % of Jordan's exports were destined to the EU. Similar disparities exist in terms of imports from the EU, although the differences are lower. This reveals the importance of the EU as a trade partner for specific countries and also the heterogeneity within the MPCs.

Since 2000 the EU's share of the MPCs' trade has reduced slightly, with the exception of Tunisia which is buying more from the EU markets and this despite the ratification of the Barcelona Agreement by a number of MPCs (discussed in the next section). Nevertheless, between 2001 and 2005, the total volume of imports from the EU by the MPCs rose from 64.6 to 88.8 billion Euro and the total volume of exports to the EU rose by about 20 billion Euro (from exports of 82.8 billion Euro to 101.8 billion Euro) (EUROSTAT, 2006).

Figure 2.3: Share of the EU-25 in the external trade of MPCs in %, 2000 and 2004

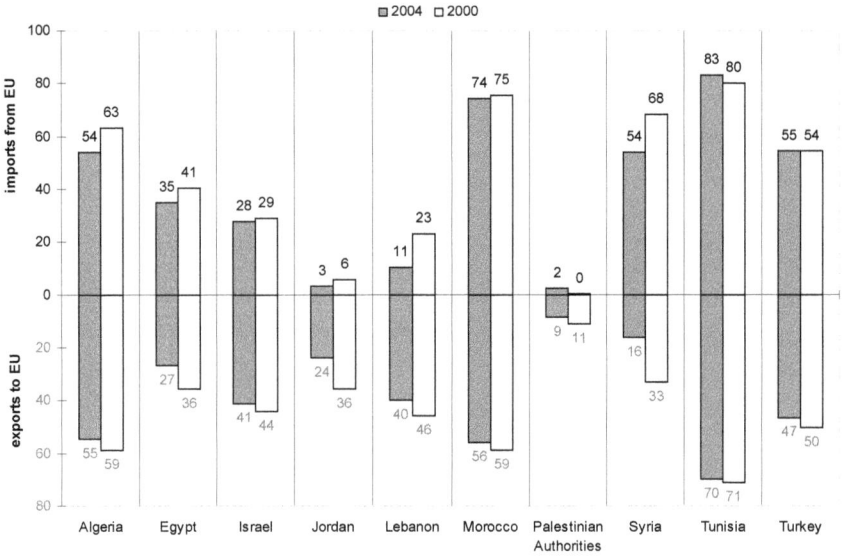

Source: Own compilation based on EUROSTAT, 2006

Food and agricultural products represent only a small proportion of the trade between the EU and the MPCs. In 2005 agricultural products, food and tobacco accounted for about 6.7 % of total imports into MPCs from the EU-25 and 4.6 % of the total exports from MPCs into the EU-25. This is equivalent to 9.5 % of the total extra-EU imports of agricultural commodities and 8.8 % of extra-EU exports. The share of agricultural commodities within the overall trade between the EU and the MPCs decreased between 1995 and 2000 (in 1995 it was round 10 %), but has remained quite stable since 2005 (QUEFELEC, 2004; EUROSTAT, various years).

The EU is the main importer of agricultural commodities from the MPCs (GALANOPOULOS et al., 2007). More than half of the agricultural exports from MPCs are destined to EU markets and a significant share of their agricultural imports also originate from the EU, although these figures vary from country to country (QUEFELEC, 2004; AQUILA and VELAZQUEZ, 2002). For example in 2004, around 70 % of

Algeria's exports of agricultural commodities were into EU markets, while for Jordan this figure was just 1.2 %. Between 30 % and 40 % of North African MPCs' purchases of agricultural commodities are from EU countries, while for Middle Eastern MPCs (i.e. Jordan and Israel) the EU's share is about 15 % (EUROSTAT, 2006).

The MPCs mostly export fresh and processed fruit and vegetables to the EU and import cereals, dairy products and sugar. In 2004 imports of fresh fruit from the MPCs accounted for about 30 % of the total imports of agricultural commodities into the EU-25, excluding fish and other seafood. Once the imports of fresh vegetables and processed fruits and vegetables are added to this figure, then more than 60 % of fruits and vegetables coming into the EU are from the MPCs (Table 2.1). Most important among these commodities are citrus fruits (especially oranges), walnuts (coming exclusively from Turkey) and tomatoes (EUROSTAT, various years; NILSSON et al., 2007).

The MPCs are the only countries from which the EU imports olive oil, tomatoes, potatoes, figs, dates, walnuts and seeds. Tunisia and Turkey are the sole suppliers of olive oil, selling almost exclusively to Italy. The specialisation of the MPCs in exporting fruits and vegetables can be explained by the comparative advantages that they possess in producing these commodities, particularly tomatoes, oranges and olive oil (GALANOPOULOS et al., 2007; NILSSON et al., 2007).

According to EUROSTAT, the MPCs mostly export their agricultural commodities to Germany, France and the Netherlands. The largest suppliers are Turkey, which between 2000 and 2005 accounted for about half of the MPCs exports of agricultural commodities to the EU, followed by Morocco, and Israel, each supplying about 20 %. The main EU countries exporting commodities to the MPCs are the central-northern EU Member States, with Germany and the Netherlands with the highest proportion of traded volumes, mostly destined to Algeria, Morocco, Turkey and Israel.

Table 2.1 shows the development of the EU's share of imports and exports of agricultural commodities to and from the Mediterranean Partner Countries. No specific trend is observable and the proportions of the trade of agricultural commodities between the EU and the MPCs between 2000 and 2005 appear to have remained unchanged.

Table 2.1: Share of MPCs in the extra EU-25 trade of agricultural commodities (HS01-24) excluding fish (HS03)

Commodity	Share in EU-25 agricultural imports from MPCs, in %		Share in EU-25 agricultural exports to MPCs, in %	
	2000	2005	2000	2005
Live animals (HS01)	0.2	0.2	4.5	3.0
Meat (HS02)	0.3	0.2	3.8	0.7
Dairy products (HS04)	0.3	0.1	15.8	14.2
Other products of animal origin (HS05)	2.4	1.8	0.4	1.0
Live plants (HS06)	4.9	2.8	0.8	1.2
Fresh vegetables (HS07)	13.9	17.6	2.7	3.0
Fresh fruit (HS08)	35.5	31.7	0.6	1.5
Coffee, tea etc. (HS09)	1.1	0.8	0.5	0.5
Cereals (HS10)	1.0	0.6	20.1	18.3
Milling products (HS11)	0.3	0.3	3.9	2.7
Oilseeds (HS12)	3.6	3.2	2.4	3.2
Lac, gums etc. (HS13)	0.7	0.5	0.4	0.8
Other vegetable products (HS14)	0.4	0.2	0.0	0.0
Other animal and vegetable fat (HS15)	5.1	8.3	8.1	4.1
Processed meat and fish (HS16)	5.4	4.5	0.9	0.9
Sugar (HS17)	1.7	2.0	10.7	10.1
Cocoa (HS18)	0.4	0.6	2.0	2.6
Processed cereals (HS19)	0.7	1.0	3.8	5.9
Processed fruits and vegetables (HS20)	14.1	16.9	1.6	2.1
Miscellaneous food (HS21)	2.3	2.2	5.2	6.5
Beverages (HS22)	1.3	1.7	3.8	5.8
Residues of food industry (HS23)	0.5	0.4	3.5	3.1
Tobacco (HS24)	3.8	2.2	4.4	8.8
All agr. commodities (HS01-24)	100	100	100	100

	2000	2005
Total value of agricultural imports excluding fish (HS03)	€4240 million	€6258 million
Total value of agricultural exports excluding fish (HS03)	€5577 million	€5351 million

Source: EUROSTAT; Own calculations

Agricultural trade in the Mediterranean basin

The structure of agricultural exports from the MPCs is very similar to that from the EU Mediterranean Member States, such as Spain, Greece and Italy, implying a high level competition among these two groups of countries (AQUILA and VELAZQUEZ, 2002). Based on 1998-99 data the authors have calculated the export similarity index (Sim_{ip}), which is given by the form: $Sim_{ip} = (\sum_j min (Q_{EUij}, Q_{EUpj}))*100$ with Q_{EUij} the j product share on i EU country agro-food exports to EU and Q_{EUpj} the j product share on p MPC's agro-food exports to EU. They found that Spain and Greece have export similarity indexes that mostly closely match those of the MPCs (46.1 and 43.6 respectively), meaning that these two countries compete the most with the MPCs. Other countries with high export similarity indexes are the Netherlands (32.6), Italy (30.9) and Portugal (27.2), while Ireland and Finland have the lowest export similarity indexes (13.9 and 12.1 respectively). Looking at the export similarities between single countries, the authors found Spain and Morocco to be the most similar (index value of 43.4) followed by Greece and Turkey (41.3), Israel and the Netherlands (37.4) and Israel and Spain (36.3). The authors point out that the high export similarity indexes are due to the leading role of fruit and vegetables and olive oil in the export portfolios of these countries. The similar export performance of the Netherlands with the MPCs could be attributed to the export of greenhouse vegetables.

In the same study AQUILA and VELAZQUEZ (2002) also examined the similarity of exports from EU countries to non-EU ones with the imports of agricultural commodities from the MPCs by calculating the complementarity index. The index helps to compare the EU export structure to non-EU countries and the MPCs import structure from EU, both for the two areas as a whole and by country and is calculated using the form: $Comip = (\sum_j min (Q_{iEXj}, Q_{EUpj}))*100$, where Q_{iEXj} the j product share on i EU country agro-food exports to non-EU countries and Q_{EUpj} the j product share on p MPC's agro-food imports from the EU. A high index value implies that the import structure of an MPC resembles the export structure of an EU Member State. The results reveal that among the MPCs the imports of Israel and Lebanon most closely mirror the exports of EU countries, with index values of 56 and 49 respectively, while the exports of Belgium, Germany, the Netherlands and France most closely resemble the imports of the MPCs (index values of 61, 58, 52 and 51 respectively). As expected the EU Mediterranean Member States have medium to low complementarity rates where Greece ranks the lowest (27).

In summary, analysis of the trade flows around the Mediterranean basin clearly shows that the EU is the most significant trade partner of the MPCs and that the MPCs are also important trade partners of the EU. Trade flows, particularly of agricultural commodities, have a stronger north-south dimension, a less strong south-north dimension, while south-south trade between the MPCs is very limited.

Clear patterns of trade in agri-food products can be seen. The EU buys fruits and vegetables and olive oil from the MPCs and sells cereals, dairy commodities and sugar to them. The agro-food export profile of the MPCs resembles that of the EU Mediterranean Member States in terms of the exported commodities and the destination markets, with their imports being complementary to the exports of central-northern EU Member States. Given this it might be in the interests of northern EU countries and the MPCs to seek stronger trade relationships around the Mediterranean basin, as both groups of countries could expand their exports but this would be against the interests of the EU Mediterranean Member States because of the high competition they face with the MPCs.

The existing trade patterns and their likely future development are largely attributed to the comparative advantage that countries have in producing certain commodities, to their geographic proximity and to historical relationships arising from trade agreements and agricultural policy, points which are discussed below.

Despite the historical relationships between the EU and the MPCs and the launch of the Euro-Mediterranean agreements, trade has not expanded over the years between the two groups of countries. This is particularly evident for fruits and vegetables, where the MPCs have not improved their market shares to the EU over the period 1995-2005, while the trade surplus of the EU felt as well. The findings of NILSSON et al. (2007) and MARTÍNEZ-GOMEZ and GARCÍA ÁLVAREZ-COQUE (2005) show that the competitiveness of the EU and the MPCs has deteriorated over the years and suggest that the trade performance of the MPCs depends up to a high grade on their favourable treatment within the Barcelona Agreement.

2.2 Trade agreements

Efforts to promote trade relationships between the EU and the MPCs have intensified in recent years and are reflected though the compilation of trade agreements with their main trade partners, as identified in the previous section.

The Euro-Med Association Agreements have been central to the evolution of trade between the EU and the MPCs. They were established in 1995, at the Summit of Barcelona and were the result of the Mediterranean Policy of the EU, which started in 1957 with the formation of the European Communities and intensified at the beginning of the '70's with the Global Mediterranean Policy (MASALA, 2000). Under this Policy the EU aimed to increase its presence in the Mediterranean Region by boosting trade and strengthening the socio-economic structures of the southern and eastern Mediterranean countries. The Barcelona Agreement was the result of this Mediterranean regional policy and declared its official aim as providing a framework for political dialogue and comprehensive cooperation among Partner Countries (EU COMMISSION, 2008a). The Barcelona Declaration (EU COMMISSION, 2008b) is not solely economic, but has in total four pillars concerning partnership in political and security, economic and financial, in social, cultural and human affairs and finally in migration. The governing body of the Euro Mediterranean Partnership is the meeting of the Ministers, hosted by the EU Member State with the EU Presidency and chaired by this country's respective Minister. The ministerial meetings are supported by the Euro-Med Committee, which consists of senior officials from EU Member States and the MPCs and by a number of technical sub-committees. In practice the EU institutions prepare the Euro-Med meetings and have, albeit unofficially, undertaken responsibility to keep the Barcelona Agreement on track (MASALA, 2000). Today the Barcelona Agreement forms a part of the European Neighbourhood Policy (ENP). Hence, the political and social dimension of the partnership is supposed to strenghen. The ENP is aimed at the EU's neighbours that have little immediate prospect of membership, but who are willing to undertake economic and particularly political reforms that extend beyond those within the current association and cooperation agreements (ERDLE, 2007).

The economic partnership is implemented through Association Agreements[3] with each of the Partner Countries, which have replaced the Cooperation Agreements of the '70's. According the EU Commission, Association Agreements have now been completed with most of the MPCs. They have been in force between the EU and Tunisia since 1998, with Israel and Morocco since 2000, Jordan and Lebanon since 2002, Egypt since 2004 and on an interim basis with the Palestinian Authorities (EU COMMISSION, 2008a). An agreement was signed with Algeria in 2001 but is still in the phase of ratification. Negotiations were initiated with Syria in 2004. Of the original Mediterranean Partner Countries, Cyprus and Malta entered the EU in 2004, while Turkey is a candidate country and connected to the EU through a customs union since December 1995. Libya is not yet a Partner Country, but is an observer. An overview of the status of the Euro-Med Agreements follows in Table 2.2. The Euro-Med Agreements have a reciprocal character and foresee north-south as well as south-south integration.

The aim of the Euro-Med Agreements is the formation of a Free Trade Area between the Partner countries after 2010, accompanied by economic and financial cooperation. The Barcelona Agreement sets out a transition period with all MPCs to gradually eliminate all tariff and non-tariff trade barriers in manufactured products. A gradual liberalisation of agricultural trade is also foreseen but with no specific time framework or provisions (PRADA LEAL and DEKA, 2004). So far the Euro-Mediterranean Agreements involve liberalisation of trade in manufactured goods and services and the establishment of a Pan-Euro-Mediterranean cumulation of origin, which was adopted by

---

[3] The Association Agreements are called Euro-Mediterranean Association Agreements or simply Euro-Med Agreements and, although not identical, this term has been used in the literature as synonymous with the Barcelona Agreement. The Barcelona Agreement is a wider term referring to all three dimensions as mentioned above, not just to the economic dimension of the Euro-Mediterranean integration process.

the Council of the EU on October 2005 (EU COMMISSION, 2008a). Financial assistance is awarded by the EU in form of grants under the MEDA programmes (Measures d'accompagnement) (Council Regulation no EC 1488/96 and its amending regulations, Official Journal of the European Communities L 189)

Table 2.2: Status of the Association Agreements between the EU and the MPCs as of January 2008

MPC	Status
Algeria	signed on April 2002 / in process of ratification
Egypt	signed on June 2001 / in force since June 2004
Israel	signed on November 1995 / in force since June 2000
Jordan	signed on November 1997 / in force since May 2002
Lebanon	signed on June 2002 /in force since April 2006
Libya	observer
Morocco	signed on February 1996/ in force since March 2000
Palestinian Authority	signed on February 1997 / in force since June 1997
Syria	initiated on October 2004 / to be signed
Tunisia	signed on June 1995 / in force since March 1998
Turkey	Customs union signed on March 1995 / in force since December 1995

Source: Own illustration adapted from EU COMMISSION, 2008a

Progress in liberalising trade in agricultural commodities has been rather limited. This has been acknowledged by the Ministerial meetings of the Barcelona Agreement and at the anniversary conference of the Barcelona Agreement in November 2005 a new 5-year work plan was developed, with a new roadmap for the liberalising agricultural trade. According to this a new round of negotiations between the EU and the MPCs started in the first quarter of 2006 (EU COMMISSION, 2006). The reason for the slow liberalisation of the agricultural sector is the sensitive nature of the sector for both sets of participants (the EU and the MPCs), since opening the markets of the MPCs to the EU and to other MPCs would result in a loss of import taxes and opening the EU markets to commodities from the MPCs would create competition problems for the Mediterranean EU Member States. AGHROUT (2007) contends that the EU's Common Agricultural Policy is an additional obstacle to the negotiations, since it shelters the EU's agricultural sector. He anticipates that some sensitive products could be excluded from liberalisation. This argument is further elaborated by the Femise Report[4] on the tenth anniversary of the Barcelona Declaration (RADWAN and

---

[4] The Femise Network is a Euro-Mediterranean non-profit association of independent economic institutes supported by the EU Commission, aiming at conducting socio-economic analysis of the Euro-Mediterranean Partnership. It currently has 75 members. Further information on the activities

REIFFERS, 2005) which shows how the EU has developed a complicated set of norms that effectively prohibit access to its agricultural markets.

The Barcelona Agreement has maintained a bilateral status so far, boosting north-south relationships. The Euro-Med Agreements seem to have a hub-and-spoke character instead of a reciprocal one and lack any institutional framework for south-south integration. GAVIN (2005) notes that the MPCs are only linked with the EU through a network of bilateral trade agreements. The author contends that this is because the Barcelona Agreement is operated by EU institutions. The economic partnership is implemented by negotiations at a bilateral level between the EU Commission and the individual MPCs. Therefore the decision process is dominated by the donor, the EU, than by a mutually agreed decision-making system. One challenge for the future of the Euro-Med Agreements and essential for the realisation of the Free Trade Area is south-south integration, which needs to be boosted by the MPCs themselves.

The southern and east Mediterranean countries are involved in a number of other bilateral and regional agreements. An overview of these and their current status in the World Trade Organisation (WTO) can be found in Table 2.3.

To start chronologically with regional agreements, three Mediterranean Partner Countries were co-founders of the Arab Maghreb Union. This Union was formed at the Summit of Marrakesh in Morocco in 1989 between Algeria, Libya, Mauritania, Morocco and Tunisia, with the main objectives of encouraging trade among its members, promoting free movement of goods and people among the Maghreb countries and incorporating a defence clause that prohibits aggression between the five countries (UMA, 2008).

The Greater Arab Free Trade Area (GAFTA) alternatively known as the Pan-Arab Free Trade Area was formed on 1st January 1998 among the countries of the Arab League. The Agreement contains provisions for the trade of goods and aims to establish a Free Trade Area that includes Bahrain, Egypt, Iraq, Jordan, Kuwait, Lebanon, Libya, Morocco, Oman, Qatar, Saudi Arabia, Sudan, Syria, Tunisia, the United Arab Emirates and Yemen (WTO notifications, 2006). In the Agreement a transition period of 10 years was foreseen for the elimination of trade duties, but Saudi Arabia communicated to the WTO secretariat that all duties and other restrictive regulations had been abolished by 1st January 2005 (WTO Secretariat, 2006). However intra-Arab integration remains relatively low. RADWAN and REIFERS (2005) identify the lack of institutions to support and even more to establish the necessary regulations for regional integration as a main reason for this, and that this in turn stems from a lack of political leadership and of the will of Arab countries to integrate.

---

of the network can be found on http://www.femise.org/index.html

A further regional agreement is the Agadir Agreement, which was signed in 2004 by Egypt, Jordan, Morocco and Tunisia. The Agreement was initiated by Morocco in 2001 after the fourth meeting of the foreign ministers of the EU and the MPCs. RADWAN and REIFFERS (2005) point out that there was no necessity for this Agreement, since the GAFTA was already in existence. However, it was welcomed by the EU as a cornerstone for promoting south-south integration. The Agreement aims to create a Free Trade Area among the signatory countries by 2006 and to harmonise standards and customs procedures. Other MPCs are welcome to enter the Agreement. One key element of the Agreement is the adoption of the Pan-Euro Mediterranean Protocol of Origin as a standard system of rules of origin. This will bring benefits from cumulation as the system is harmonised with the EU one, but this will raise cost issues for the signatory countries, which will need to change and adapt their existing system of certifying origin. This Agreement came into force in 2007 (EU COMMISSION, 2008c). RADWAN and REIFFERS (2005) argue that the Agadir Agreement can contribute more to integration among the four signatory countries than GAFTA because it is supported by the EU and its institutions. Moreover, the Agreement is among quite similar Arab Mediterranean countries (in terms of their export range and industrial structure) implying that these countries have similar interests, which could facilitate agreement. Many GAFTA members are Gulf countries, that are richer, have economies based on oil and thus have differing interests. RADWAN and REIFFERS (2005) also point out that the effect of the Agadir Agreement might be diluted by the GAFTA, which is already in force, and by the high trade reliance of Morocco and Tunisia on trade with the EU, which leaves little space for developing south-south trade. Indeed the European markets appear more interesting than those of neighbour countries to most MPCs. Finally the purpose of the agreement is not completely clear, since the four signatory countries were already involved in a Free Trade Area within the framework of GAFTA and had already concluded bilateral trade agreements with each other. Hence, the Agadir Agreement appears to complicate and confuse the trade relationships and make the respective processes more bureaucratic.

In addition the MPCs are engaged in a number of bilateral trade agreements, also shown in Table 2.3.

Table 2.3: WTO status, bilateral and regional agreements of the MPCs as in January 2008

	WTO	GAFTA	AMU	Agadir Agreement	USA	Other
Algeria	application 1987		1989			
Egypt	member 1995	1998		2004		EFTA since 2007 COMESA since 1994
Israel	member 1995				FTA 1995*	Mexico since 2000 Turkey since 1997 Canada since 1997 EFTA since 1993
Jordan	member 2000	1998		2004	FTA 2000	Singapore since 2005 EFTA since 2002
Lebanon	application 1999	1998				EFTA since 2007
Libya	application 2004	1998	1989			
Morocco	member 1995	1998	1989	2004	FTA 2004	Turkey since 2006 EFTA since 1999
Palestine	–	in progress				Turkey since 2005 EFTA since 1999
Syria	–	1998				Turkey since 2007
Tunisia	member 1995	1998	1989	2004		Turkey since 2005 EFTA since 2005
Turkey	member 1995					Syria since 2007 Morocco since 2006 Tunisia since 2005 Palestine since 2005 Croatia since 2003 Bosnia since 2003 FYROM since 2000 Israel since 1997 EFTA since 1992 ECO since 1992

Notes: * The Agreement was signed in 1985 but only entered into full force in 1995. In 2004, with the agreement on agriculture the FTA was expanded to include agricultural commodities too.

COMESA (Common Market for Eastern and South Africa): Angola, Burundi, Comoros, Democratic Republic of Congo, Djibouti, Egypt, Eritrea, Ethiopia, Kenya, Madagascar, Malawi, Mauritius, Namibia, Rwanda, Seychelles, Sudan, Swaziland, Uganda, Zambia and Zimbabwe); ECO (Economic Cooperation Organisation): Afghanistan, Azerbaijan, Iran, Kazakhstan, Kyrgyz Republic, Pakistan, Tajikistan, Turkey, Turkmenistan, Uzbekistan); EFTA (European Free Trade Association countries): Iceland, Liechtenstein, Norway, Switzerland; FTA: Free Trade Area; FYROM: Former Yugoslavic Republic of Macedonia

Source: Own compilation based on WTO notifications, 2008

The USA has signed Free Trade Area Agreements with Israel (in force since 1995), Jordan (2000) and Morocco (2004). These agreements are expected to boost US exports to the partner countries in absolute terms, but their overall effect on US exports, production and the economy is considered to be rather negligible due to relatively low values of US exports to Israel, Jordan and Morocco (BUTCHER et al., 2000 and 2004). Although these FTAs do not have any profound effect for the USA, free market access to Mediterranean countries and especially to partner countries of the EU increases the competitiveness of the US producers (manufactures and farmers) not only against local producers but also in relation to EU ones.

In addition most of the MPCs have already introduced a Free Trade Agreement with the European Free Trade Association countries (EFTA) according to the WTO notifications (2008), since they are connected to the EU with Association Agreements and have already started to form bilateral trade agreements with each other.

Turkey has been particularly active in concluding FTA agreements with other MPCs. This might be partly attributed to Turkey being an EU candidate country and therefore needing to conclude and apply bilateral trade agreements with the EU's trade partners. Thus, Turkey has negotiated FTA agreements that are in force with Israel since 1997, with Morocco since 2006, with the Palestinian Authorities since 2005, with Syria since 2007 and with Tunisia since 2005. Turkey is also member of the Economic Cooperation Organisation (ECO), a Free Trade Area among Afghanistan, Azerbaijan, Iran, Kazakhstan, the Kyrgyz Republic, Pakistan, Tajikistan, Turkey, Turkmenistan and Uzbekistan.

The only MPCs that have other bilateral FTA agreements than with the USA, the EFTA countries and Turkey are Israel, Jordan and Egypt. Israel concluded an FTA with Canada in 1997 and with Mexico in 2000, while Jordan has one with Singapore since 2005. Egypt is member of the Common Market for Eastern and Southern Africa (COMESA), a Free Trade Area among African countries that entered into force in 1994.

The MPCs have shown an interest in participating in the current WTO negotiations - most of them are members of the WTO or have applied for a membership. Despite the fact that the MPCs are involved with a number of regional trade agreements, they appear to have different interests and certainly they have not adopted a common strategy in the Doha round of negotiations.

GARCIA ÁLVAREZ-COQUE (2006a) notes that all the MPCs have moved towards implementing the Agreement on Agriculture and have committed themselves to reducing export subsidies, domestic support and import duties on agricultural products. However, they have followed different approaches and have developed different policies for integrating their agricultural sector into a multilateral trade environment. He points out that Israel uses all three of these pillars to support its agriculture, while all other countries have opted for a more limited combination of farm support options.

Turkey is the only MPC entitled to grant export subsidies to its producers, but this is limited by special safeguards and trade-distorting domestic support beyond the *de minimis* level (10 % of the production value for developing countries). Tunisia and Morocco have the right to special safeguards for 32 and 374 commodities respectively, but are not allowed to provide export subsidies. Jordan has kept an option on trade-distorting domestic support but is not invoking export subsidies or special safeguards. Egypt has no rights beyond the *de minimis* trade-distorting support (GARCIA ÁLVAREZ-COQUE, 2006a).

The different approaches of the MPCs towards multilateralism are visible through the various interest groups that they align themselves in the WTO negotiations (GARCIA ÁLVAREZ-COQUE, 2006a). Israel is member of the G-10 group (net importers) whose main goal is non-trade concerns, Egypt is part of a group seeking a reduction of farm subsidies in industrial countries and less focus on market access (the G-20, which includes major developing countries as India, Brazil and China). Turkey participates in the G-33 (vulnerable developing economies) which requested special treatment for particular commodities and support to small farmers. Finally Morocco, Tunisia and Egypt are members of the G-90 (consisting of African and least developed countries and those of the Asian-Caribbean-Pacific group) that seek to preserve their preferential treatment, fearing the impacts of preference erosion.

Finally it is worth noting that some MPCs are members of regional trade agreements with a partial scope. Hence, Algeria, Egypt, Libya, Morocco and Tunisia are members of the General System of Trade Preferences among Developing Countries (GSTP), an agreement that entered into force in 1989. Egypt, Israel, Tunisia and Turkey are co-signatories of the Protocol relating to Trade Negotiations among Developing Countries (PTN), which has been in force since 1973, and five years earlier, in 1968, Egypt signed the Tripartite Agreement together with India and the former Yugoslavia.

2.3 Development of trade protection among the Mediterranean countries

Despite the Euro Med Agreements and the numerous preferential trade agreements among the countries surrounding the Mediterranean, there are still restrictions on trade in agricultural commodities and obstacles to liberalisation and the creation of a Free Trade Area. Trade protection is expressed through both import restrictions and export subsidies. Table 2.4 provides an overview of the import restrictions imposed by the EU on commodities coming from Mediterranean countries, expressed as *ad-valorem* equivalents.

Table 2.4: Weighted averages of ad-valorem equivalents of applied tariffs by the EU from imports from the MPCs in 2005

Commodity	Algeria	Egypt	Israel	Jordan	Lebanon
Bovine meat	n.a.	n.a.	n.a.	20 (1996)	n.a.
Pig meat	n.a.	0	0 (2003)	0 (1996)	0 (2003)
Poultry meat	n.a.	n.a.	3.15	8.8 (1996)	8.8 (1996)
Milk and cream not concentrated	n.a.	n.a.	n.a.	n.a.	n.a.
Milk and cream concentrated	n.a.	n.a.	n.a.	n.a.	n.a.
Tomatoes	7.6	2.64	10.9	3.87	4.38
Oranges	0.63	4.17	5.84	14.86	4.15
Apples	2.9	2.9	2.83	n.a.	6.1 (2001)
Wheat and meslin	0 (1996)	12.8 (2001)	2.13	n.a.	0
Barley	n.a.	n.a.	n.a.	n.a.	n.a.
Maize (corn)	n.a.	0	0	0 (1994)	n.a.
Rice	n.a.	7.7	n.a.	n.a.	n.a.
Grain sorghum	n.a.	6.4	8.8 (1997)	n.a.	n.a.
Other cereals	n.a.	n.a.	n.a.	n.a.	n.a.
Soya beans	0 (1996)	0 (1999)	0	n.a.	0 (2001)
Linseed	n.a.	0	n.a.	n.a.	n.a.
Sunflower seeds	0 (1996)	0	0	0 (1997)	0 (1995)
Other oil seeds	n.a.	0	0	0 (2003)	0
Soya-bean oil	n.a.	n.a.	7.35	n.a.	4.05 (2001)
Olive oil	n.a.	n.a.	n.a.	n.a.	0
Sunflower, safflower, cottonseed oil	n.a.	4.6	n.a.	n.a.	5.61 (1997)
Unprocessed tobacco	n.a.	n.a.	n.a.	n.a.	0
Cotton not carded or combed	0 (1996)	0	0	0 (1995)	0 (1995)
Cane sugar	n.a.	n.a.	n.a.	n.a.	n.a.
Beet sugar	n.a.	n.a.	n.a.	n.a.	n.a.

Agricultural trade in the Mediterranean basin

Table 2.4 – continued –

Commodity	Libya	Morocco	Syria	Tunisia	Turkey
Bovine meat	n.a.	n.a.	n.a.	n.a.	n.a.
Pig meat	n.a.	0 (2003)	n.a.	n.a.	0 (1996)
Poultry meat	n.a.	2.9	4.4 (2001)	n.a.	0
Milk and cream not concentrated	n.a.	n.a.	n.a.	n.a.	n.a.
Milk and cream concentrated	n.a.	n.a.	n.a.	n.a.	n.a.
Tomatoes	10.9	9.81	10.9	0.51	0
Oranges	14.86	2.55	11.36	1.81	0
Apples		2.9	2.9	6.3 (1999)	0
Wheat and meslin	n.a.	20 (1994)	0	n.a.	0
Barley		n.a.	n.a.	n.a.	n.a.
Maize (corn)	n.a.	0	n.a.	0 (2001)	0
Rice	n.a.	n.a.	n.a.	0	0
Grain sorghum	n.a.	n.a.	n.a.	7.6 (1999)	n.a.
Other cereals	n.a.	n.a.	n.a.	n.a.	n.a.
Soya beans	n.a.	n.a.	n.a.	0	0
Linseed	n.a.	0	n.a.	n.a.	0
Sunflower seeds	n.a.	0 (1999)	0	n.a.	0
Other oil seeds	0 (1993)	0	0	0	0
Soya-bean oil	n.a.	3.85	n.a.	4.05 (2001)	0
Olive oil	n.a.	n.a.	n.a.	n.a.	n.a.
Sunflower, safflower, cottonseed oil	n.a.	2.44	4.6	2.54	0
Unprocessed tobacco	n.a.	n.a.	n.a.	n.a.	0
Cotton not carded or combed	n.a.	0	0	0	0
Cane sugar	n.a.	n.a.	n.a.	n.a.	n.a.
Beet sugar	n.a.	n.a.	n.a.	n.a.	n.a.

Notes: All data for 2005 unless indicated otherwise in parenthesis; n.a.: not available
Source: Own compilation; TRAINS

The EU imposes entry prices for specific volumes of imports of fruits and vegetables and has an import quota for olive oil. Nevertheless, when the imported quantity is within the quota or within the specified quantity for which the entry price applies, then no additional tariff (or tariff equivalent) is applied. When for oranges, apples, tomatoes and olive oil no duty is reported, it is implied that the imported quantity did not exceed the specified quantity.

The table shows that the EU applies quite low tariffs on agricultural commodities imported from the MPCs. The highest tariffs are applied to imports of oranges and tomatoes, and these rates vary between individual countries. The weighted average of *ad-valorem* equivalent (AVE) of all applied tariffs is highest for imported oranges from Jordan (14.86 % in 2005), and lowest for oranges from Algeria (0.63 % in 2005). For tomatoes the respective AVE is highest for Israel, Libya and Syria (10.9 % for each country), and lowest for Tunisia (0.51 %). For the remaining commodities the tariffs are all below 10 %, with the exception of beef from Jordan (20 % in 1996) and wheat from Morocco (20 %).

In all cases the tariffs applied by the EU are lower than the Most Favoured Nation (MFN) tariff, which shows that the EU gives additional preferential treatment to its Mediterranean Partner Countries. At first sight this appears to contradict the claim, made above, that slow progress has been made in extending the preferences to agricultural commodities. Hence, before speculating whether the preferences have deepened since the Barcelona Agreement and whether this has determined the traded volumes between the EU and the MPCs, it is worth having a closer look at the trade preferences.

An indication of the evolution of trade preferences in the Mediterranean basin granted by the countries involved in this agreement could be given by the value of the preference margin (VPM) as an indicator of the economic value of trade preferences. GRETHE et al. (2006)[5] argue that the VPM of all agricultural commodities for all MPCs covered by the agreements of the mid-70s was about 130 million Euro, whereas in 1995 the VPM was about 190 million Euro (an increase of 48%). After the Barcelona Agreement this reduced to about 165 million Euro. According to the authors this negative change is attributed to a reduction in EU MFN tariffs. They argue that once all Euro-Med Agreements have entered into force the VPM will reach 226 million Euro.

Tables A.1 and A.2 (see Annex) present details of the VPM from imports into the EU of selected agricultural commodities between 1999 and 2003[6]. The calculations were done following GRETHE and TANGERMANN (1998a) i.e. it has been assumed that both preferential and non-preferential commodities are sold in the destination market (the EU) at the same price and thus that the value of the preference margin is the price difference between preferential and non-preferential exports multiplied by the quantity of the commodity each partner country exported into the EU. A thorough presen-

---

[5] This study has been also published as GRETHE et al. 2006 and can be found as GRETHE et al. 2005a and b. An earlier version of this study has been published as a discussions paper of the University of Göttingen (GRETHE and TANGERMANN, 1998b).

[6] These commodities are the ones included in the database of the model AGRISIM, which is used further in this study for the own empirical analysis. The time period was chosen based on data availability.

tation of the way the VPM has been calculated follows in the Annex. In most cases the MFN duties are applied and thus the VPM is zero. MPCs gain due to the preferential treatment only for their main export products, such as fruits and vegetables. The size of the VPM for a given commodity differs significantly from country to country, mainly because of the high variation in quantities exported and not because of any variation in the preferential duty compared to the MFN one. This said the difference between the MFN and the applied duty varies between 0.2 and 7 %.

A comparison of the VPM of 2003 with that of 1999 clearly shows that the Barcelona Agreement has only slightly intensified the benefits for the MPCs (Figure 2.4). In total the VPM increased by about US$ 3 billion throughout the MPCs and for selected agricultural commodities, but this increase is because of Morocco, where the VPM from about US$33.6 million in 1993 expanded to almost US$102.3 million. A potential expansion of exports into the EU of those commodities where the VPM is already positive would result in significant gains for the MPCs.

Figure 2.4: Value of Preference Margin for selected agricultural commodities resulting from the Euro-Mediterranean Association Agreements

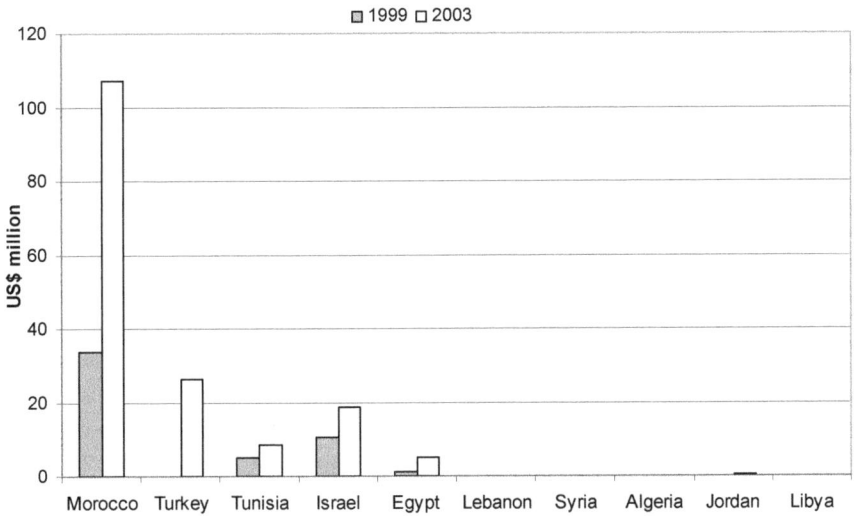

Source: Own calculations based on reported import duties derived from TRAINS and bilateral trade flows derived from COMTRADE

It seems therefore that the MPCs expect to gain from the Barcelona Agreement more from trade diversion effects than from trade creation. Moreover, because they compete with the EU Mediterranean Member States and produce at lower cost, it is expected that a complete trade liberalisation could divert trade from the Mediterranean

EU countries to the MPCs when it comes to trade with the northern EU Member States.This is feared by many Mediterranean EU countries (and their producers) and is a reason for resisting the full implementation of the Euro-Med Association Agreements (KAVALLARI et al., 2005a).

Table 2.5 lists the *ad-valorem* equivalents of import restrictions imposed by the MPCs on commodities coming from the EU. Due to the EU's high share of total imports into the MPCs, the import tariffs upon EU commodities are of particular relevance for the MPCs and are a main source of tariff revenues. It is clear that the MPCs apply much higher tariffs to imports from the EU than the EU does on imports from the MPCs. The entries refer either to year 2005 or to the latest available year after the conclusion of the Euro-Med Agreements.

The highest tariff rates are applied to livestock commodities. The highest levels of the AVE (weighted average) were applied by Morocco to EU beef (272 % in 2003), followed by Tunisia on imported non-concentrated milk (180 % in 2004) and by Turkey on both concentrated and non-concentrated milk (150 % in 2003). In crop products the AVE of import tariffs (as a weighted average) is lower and only exceeds 100 % in the case of tomatoes imported into Tunisia (165 % in 2003), sugar (both cane and beet) imported into Turkey (136.5 % in 2003) and EU olive oil imported into Israel (127.5 % in 2005).

Morocco, Turkey and Tunisia seem to apply the highest tariffs on agricultural commodities from the EU. The southern Mediterranean countries (e.g. Algeria, Egypt, Libya, Morocco and Tunisia) appear to levy higher rates than those of eastern Mediterranean (e.g. Jordan, Lebanon and Syria), and this may be related to differences in the trade volumes and the EU's share of these countries total imports. As shown earlier in this chapter, EU commodities account for a lower share of the total imports of eastern Mediterranean countries, but account for more than half of the total imports of southern Mediterranean countries. Hence, for Jordan, Lebanon or Syria it appears less attractive to apply high import tariffs to EU commodities, wheras for countries such as Morocco, Tunisia or Algeria the high protection both brings in tax revenue and can be justified as a means of protecting domestic production.

The TRAINS database only reports on tarrifs for a limited number of commodities imported into the MPCs. These are at about the same level as the duties for EU commodities and the applied tariffs are as high as the MFN tariffs (Table A.3 in the Annex).

In general the MPCs have not granted preferences to the EU or to other MPCs for agricultural commodities. Despite the reciprocal character of the Euro-Med Agreements and the numerous preferential agreements among the MPCs, the TRAINS database reports that up to 2005 the tariffs applied to commodities imported from the EU or from other MPCs are the Most Favourite Nation tariffs.

Table 2.5: Weighted averages of ad-valorem equivalents of applied tariffs by the MPCs to imports from the EU in 2005

Commodity	Algeria	Egypt	Israel	Jordan	Lebanon
Bovine meat	40 (1993)	5 (2002)	n.a.	5 (2001)	5
Pig meat	30 (2003)	40 (2002)	n.a.	30 (2003)	5
Poultry meat	30 (2003)	80 (1998)	n.a.	17.51 (2003)	5.58
Milk and cream not concentrated	30 (2003)	25 (2002)	0 (1993)	30 (2003)	37.5
Milk and cream concentrated	5 (2003)	11.88 (2002)	57.26	3.33 (2003)	6.75
Tomatoes	30 (2002)	n.a.	0 (1993)	n.a.	70 (2002)
Oranges	30 (2003)	n.a.	n.a.	n.a.	n.a.
Apples	30 (2003)	40 (2002)	0	30 (2003)	70
Wheat and meslin	2.5 (2003)	1 (2002)	25	0 (2003)	0
Barley	10 (2003)	5 (2002)	0	0 (2001)	0
Maize (corn)	5 (2003)	1 (2002)	0	n.a.	4.68
Rice	5 (2003)	20 (2002)	0	5 (2003)	5
Grain sorghum	10 (2002)	5 (2002)	0	n.a.	n.a.
Other cereals	n.a.	n.a.	n.a.	n.a.	n.a.
Soya beans	5 (2003)	1 (1998)	0	n.a.	0
Linseed	5 (2003)	1 (2002)	0	n.a.	0
Sunflower seeds	5 (2003)	1 (2002)	0	15 (2003)	0
Other oil seeds	5 (2003)	2.43 (2002)	0.06	n.a.	0.02
Soya-bean oil	10.44 (2003)	10.02 (2002)	4	15 (2003)	8.78
Olive oil	30 (2003)	12.5 (2002)	127.5	n.a.	70
Sunflower, safflower, cottonseed oil	11.89 (2003)	8.67 (2002)	4	0.29 (2003)	15
Unprocessed tobacco	15 (2003)	n.a.	0	18.92 (2003)	0 (2002)
Cotton not carded or combed	5 (2003)	5 (2002)	0	n.a.	0
Cane sugar	5 (2003)	7.5 (2002)	0	n.a.	5
Beet sugar	5 (2002)	7.5 (2002)	0	n.a.	5

Table 2.5 – continued –

Commodity	Libya	Morocco	Syria	Tunisia	Turkey
Bovine meat	0 (1996)	275 (2003)	n.a.	73 (2004)	15 (1995)
Pig meat	n.a.	53.5 (2002)	n.a.	125 (2003)	20 (1995)
Poultry meat	50 (2002)	111.29 (2003)	n.a.	73 (2004)	64.42 (1997)
Milk and cream not concentrated	0 (2002)	109 (2003)	n.a.	180 (2004)	150 (2003)
Milk and cream concentrated	0 (2002)	99.87 (2003)	7 (2002)	58.42 (2004)	150 (2003)
Tomatoes	50 (2002)	n.a.	n.a.	165 (2003)	49.1 (2003)
Oranges	30 (2002)	n.a.	n.a.	n.a.	54.6 (2002)
Apples	30 (2002)	52 (2003)	n.a.	200	60.9 (2003)
Wheat and meslin	15 (2002)	60 (2003)	n.a.	65.79 (2003)	13.35 (2003)
Barley	0 (2002)	27.12 (2003)	n.a.	73 (2004)	42.5 (2003)
Maize (corn)	0 (2002)	75.10 (2003)	n.a.	0 (2004)	12.66 (2003)
Rice	15 (2002)	62.66 (2003)	7 (2002)	34.79 (2004)	31.33 (2003)
Grain sorghum	0 (1996)	n.a.	n.a.	34.33 (2002)	30 (1999)
Other cereals	n.a.	n.a.	n.a.	n.a.	n.a.
Soya beans	25 (1996)	9.17 (2003)	n.a.	15 (2003)	0 (2003)
Linseed	n.a.	21.25 (2003)	n.a.	43 (2003)	12 (2003)
Sunflower seeds	n.a.	12.33 (2003)	7 (2002)	43 (1998)	8 (2003)
Other oil seeds	10.34 (2002)	22.02 (2003)	n.a.	25 (1995)	4.7 (2003)
Soya-bean oil	15 (2002)	2.59 (2003)	n.a.	15.01 (2004)	25.01 (2003)
Olive oil	15 (2002)	52 (2003)	n.a.	115 (2004)	32 (2003)
Sunflower, safflower, cottonseed oil	0 (2002)	2.53 (2003)	7 (2002)	16.58 (2004)	32.23 (2003)
Unprocessed tobacco	n.a.	17.5 (2003)	n.a.	22 (2004)	25 (2003)
Cotton not carded or combed	0 (1996)	2.5 (2003)	n.a.	0 (2004)	0 (2003)
Cane sugar	n.a.	35 (2001)	n.a.	15 (1998)	136.5 (2003)
Beet sugar	n.a.	35 (2003)	n.a.	n.a.	136.5 (2003)

Notes: Unless 2005 the latest available year is indicated in the parenthesis; n.a.: not available
Source: Own compilation; TRAINS

At the multilateral level, an indication of overall trade protection can be given by the Nominal Protection Rate (NPR). The rate of nominal protection is a coefficient that shows the ratio of the domestic price to the border price, where the border price is the price of the international market converted into local currency using an exchange rate benchmark (TSAKOK, 1990, pp. 55-56). Figure 2.5 illustrates the average NPR for the period 2000-2002 and all the countries surrounding the Mediterranean. For the EU the NPR is derived from the database on Producers' Support Estimate (PSE) (OECD, 2006), while for the MPCs it is calculated by extracting the *ad-valorem* tariffs from export subsidies.

It is clear that Morocco applies higher protection rates for livestock commodities and wheat than the EU, with the exception of beef and that Turkey is the most protective country. All Mediterranean countries protect their olive oil markets, Morocco applies protection to the apple market and the other MPCs (except Morocco and Turkey) protect their tomato and orange markets. Nevertheless, because of the low imports of olive oil, fruit and vegetables imported into the MPCs, this trade protection has a relatively moderate effect on these commodities. The most significant protection is levied on livestock commodities and cereals which are the commodities imported most by the MPCs.

Figure 2.5: Nominal protection rate in %, average of 2000-2002

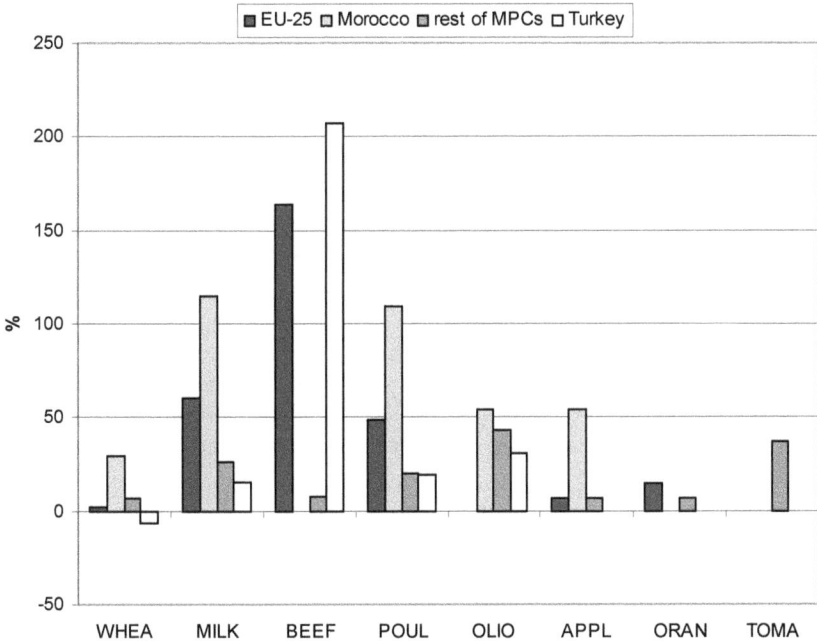

Source: Own calculations based on TRAINS; OECD, 2006

Several Mediterranean countries have applied export subsidies to a number of their commodities, particularly fruits and vegetables. Between 1997 and 2003, Israel, Morocco, Tunisia and Turkey all reported export subsidies to the WTO for fruits and vegetables. The value of these subsidies and the quantities subsidised are shown in detail in Table A.4 in the Annex.

Customs duties such as import tariffs and export subsidies are only the one side of trade protection. The other side is trade restrictions through so-called non tariff barriers (NTBs), such as technical regulations, monopolistic measures and quantity control measures. The TRAINS database reports the existence of non tariff barriers on agricultural commodities by the EU, Algeria, Egypt, Jordan, Lebanon, Morocco and Tunisia in recent years. The reported measures concern livestock commodities, live animals and fruit and vegetables and are related to the protection of human and animal health. Packaging and labelling requirements and pre-shipment inspections have also been applied to fruits and vegetables (also to protect human health) and the Lebanon has applied seasonal quotas.

Although it is difficult to quantify the impact of NTBs on import levels, it is obvious that as tariffs are reduced, NTBs become the predominant instrument to free trade. Moreover they do not have only a prohibitive character, but also a binding one. An attempt to estimate the effect of NTBs on imports has been made by KEE et al. (2006). The authors considered core NTBs (i.e. price and quantity control measures, technical regulations and monopolistic measures as single-channel imports) and agricultural domestic support to estimate the impact of non-tariff measures on imports and further to transform the quantity impacts into price changes by producing AVE of NTBs. Their results suggest that the NTBs increase world protection by about 10 %, that the most restrictive NTBs are those on agricultural commodities and that the absolute value and the relative importance of the AVE of NTBs increases with GDP per capita. Algeria and Morocco are among the five countries with the highest AVEs of core NTBs and the EU has the highest AVEs due to domestic support. This shows that the issue of NTBs is of particular relevance for Mediterranean countries. Even if trade in agricultural commodities becomes free within the Euro-Med Agreements, NTBs will certainly constrain free trade within the Mediterranean and should certainly be on the agenda of negotiations between the EU and the MPCs and between MPCs themselves.

2.4 Reforms of agricultural policies that affect trade

One of the main challenges for agriculture around the Mediterranean is hence the reforms of agricultural trade, forthcoming or just concluded. These reforms can be seen from three perspectives. The first is reforms in domestic policies that could potentially affect trade, such as, the reform of the agricultural policy of major trade players, like the EU. The second aspect is the anticipated changes in agricultural trade policy foreseen by the Euro-Med Agreements and finally the third type is changes in the agricultural trade policy at the multilateral level.

The existing Euro-Med Agreements were presented in section 2 of this chapter, which also elaborated the position of the MPCs in the current WTO round. Extending the Euro-Med agreements to agricultural commodities would mean the EU giving free market access to the commodities of the MPCs and vice versa. This could be promising for the MPCs as it could give them more export opportunities for their commodities and allow them to benefit from trade creation and trade diversion from third countries, or even of EU Member States that produce at higher cost. It would also be attractive for EU Member States that are net exporters to the MPCs, such as the northern EU Member States, who could expand their trading activities. The EU Mediterranean Member States are opposed to this, due to their fear of increased competition (as explained above).

Continuing with multilateral opening of markets, both for the EU and for the MPCs it is connected to the issue of preference erosion and thus could appear less attractive for both country groups. For example a reduction of the import tariffs for highly protected commodities, such as beef or wheat, would result in lower import rents, and the further impacts of these on production, trade flows and overall welfare would need to be systematically analysed.

The following paragraphs elaborate the latest reform of the EU's Common Agricultural Policy (CAP) and the features of this that are of most interest for Mediterranean agriculture.

The Common Agricultural Policy (CAP) has been subject to an immense reform process in recent years. The latest reform of the CAP was first discussed in 1999 by the Council of the Ministers of Agriculture in Berlin. The proposals of this Council, known as "Agenda 2000" were set to run to 2006 and were meant to be revised in 2003. The Commission published the Mid Term Review of the CAP in July 2002, as a joint communication of the Commission and the European Council. This was not just a review of the situation in agriculture, but also involved new reform proposals. These proposals were revised and adopted in 2003 by the Council of Luxembourg, now known as the Luxembourg Agreement, providing the framework for agriculture in the European Union for the next 10 years.

The reform of the Common Market Organizations (CMO) for olive oil and tobacco and the support scheme of cotton are considered as follow-ups of the Luxembourg Agreement, and came into force in the beginning of 2006. The objective is to provide a long-term future for these three sectors by promoting competitiveness, market–orientation and stable income for farmers. The changes are based on the already existing regimes for these sectors (Council of the European Union, 2004 and Council Regulation 864/2004-Official Journal of the EU L161).

The organization of the raw tobacco market was set out in Council Regulation 2075/92 and the rules of implementation rules were specified in Commission Regulation 2848/98 (Official Journal of the European Communities L 215 and L 358 respectively). The producers received a premium subject to their production quotas, which are allocated to them according to a guarantee threshold. The premium had three parts: A specific aid (25% of the premium), a variable part, adjusted by the varieties group and Member State and a fixed part, which was the difference between the premium after the deduction of the amount withheld for financing the Tobacco Fund and the sum of the specific aid and the variable part. The Community levied zero duty on imports from ACP countries and developed countries in the SPG system.

Council Regulation 1638/98 applied to olive oil (Official Journal of the European Communities L 210). It set out a production target price and thus determines the production aid received by the olive growers. This was granted on the basis of the quantity of olive oil that they actually produced. To control the budget expenditure of the Community, the aid was allocated according to National Guaranteed Quantities (NGQ) and in the case of lower or higher production of these NGQs a stabiliser mechanism was applied. The aid was distributed to the Producer Organisations and only producers registered as olive cultivators were eligible. No aid was provided for additional areas planted after 1 May 1998, with the exception of replanting old olive plantations or new plantings covered by programmes approved by the Community. The areas planted and the olive cultivation register were based on data created by the Geographical Information System (GIS) and should correspond to this data. Almost all Community imports of olive oil come from Tunisia, which had an import quota of 53,000 tonnes (in 2002) rising to 56,000 tonnes (since 2006). This quota is expected to be abolished after 2010 with the implementation of the Euro-Mediterranean Free Trade Area.

Production aid for cotton is governed by Council Regulation 1051/2001 (Official Journal of the European Communities L 148). The producers receive a subsidy per tonne of non-ginned cotton, equal to the difference between the world market price and the guide price. The later is set by the Council and the former is determined by the Commission based on prices on the international cotton fibre market. The aid is paid to the ginners, who are only entitled to receive it so long as they provide a minimum of 95 % of the guide price to the producers. This support system is applied for a Maximal Guaranteed Quantity in the Community, which is divided to National Guaranteed Quantities and is subject to a stabiliser mechanism which can be used to reduce the guide price, and thus the minimum price in case of excess production. As far as relationships with other countries are concerned, there are no custom duties, import quotas or any support measures for exports.

These support mechanisms to producers are examples of deficiency payments, as classified in the Amber Box, in the non-exempted direct payments category (as opposed to the price support category) and therefore need to be reconsidered. Given that no price cuts were necessary, the Commission's proposals only concerned the decoupling of these direct payments and the introduction of a Single Farm Payment.

The decoupled payments are linked to environmental and food safety standards through cross-compliance and are subject to the modulation and financial discipline mechanisms. Different approaches to this have been undertaken for each of these sectors, since they face different problems and there are differences in their long-term priorities (Commission of the European Communities, 2003a and b).

For olive oil it has been decided that a conversion will imply a minimum of 60 % of the coupled payments for the reference period being transferred to the single farm payment entitlements. The four marketing years 1999/00-2002/03 serve as the reference period for the calculation of the initial payments. 40 % of the initial payments may be retained by the member states as an additional acreage- based olive payment, paid through the form of a national envelope. Current private storage measures will be kept as a safety net mechanism. Refunds for exports and for manufacturing certain preserved foods will be abolished (Council of the European Union, 2004).

For tobacco the Council has decided on a gradual decoupling of the existing tobacco premium, paralleled by the establishment of a financial restructuring envelope, under the second pillar of the CAP, to support a more sustainable future for the sector. A transitional period towards full decoupling is suggested from 2006 to 2010. During this period 40 % of the current payment must be decoupled and up to 60 % can be maintained as coupled. The production quotas are being kept so as to define the volumes entitled to receive the coupled payment. At the end of this period support for tobacco production will be fully decoupled, with 50 % of it included in the single farm payment and 50 % transferred to the restructuring envelope, after which the new CMO will apply (Council of the European Union, 2004).

For cotton two types of payment are being introduced: the single farm payment and a payment of eligible acreage of cotton, representing the decoupled and coupled part of the payment respectively. Member states must transfer 65% of the support expenditure paid to producers to the single farm payment and pay the other 35% as an area payment. Greece has 370 000 ha of eligible area and will introduce a two tier structure of coupled aid (594 €/ha for the first 300,000 ha and 342.85 €/ha for the remaining 70,000 ha). Spain has 85,000 ha of eligible land (where the coupled aid will be 898 €/ha) and Portugal has 360 ha with 556 €/ha as the coupled payment. This area payment will be reduced proportionately if production exceeds the maximum within each of these member states (Council of the European Union, 2004).

Production for these three products is highly concentrated in certain regions, many of them lagging behind in economic development and employing a high proportion of the rural population. For example cotton is cultivated mainly in Greece in Thessaly, Macedonia-Thrace and Sterea Ellada and in Spain it is cultivated in Andalusia, Murcia and Valencia (Directorate-General for Agriculture, 2003). The main production areas of tobacco are the provinces of Macedonia-Thrace, Thessaly and west Sterea

Ellada in Greece, Extremadura in Spain and Umbria, Campania, Aquitane and Veneto in Italy (Commission, 2003c). Because of geographical constraints these areas do not offer many alternatives for other economic activities or for cultivating other possibly more competitive crops. Therefore, special consideration should be given to the impacts of decoupling payments on these sectors (and regions). Abandonment of production due to decoupling would generate significant negative impacts for rural development in these areas.

From the above it becomes obvious that agricultural trade relationships round the Mediterranean are characterised by complexity. Despite the several trade agreements, which are in some cases overlapping, the trade flows are built on the north-south axis and are dominated by the presence of the EU. Trade relationships seem to have so far a hub-and-spoke character, with the EU being in the centre of the cyclus and this despite the commitments of all countries to base their relationships on reciprocity.

The analysis of the preference margins and the calculations of the value of the preference margin showed that only the EU has granted preferences within the framework of the Barcelona Agreement. The size of the VPM differs significantly from country to country reflecting the bilateral status of the Barcelona Agreement. The Mediterranean Partner Countries are on the one side interested in deepening the preferences they currently enjoy, but on the other side they do not seem willing to open their borders to the EU commodities and keep high protection rates. The nominal protection rate is particularly high for those commodities that the MPCs import mainly from the EU.

To this complexity is added uncertainty about the future of the agricultural trade round the Mediterranean caused by recently agreed and by the forthcoming reforms of agricultural policies. The reform of the EU's Common Agricultural Policy makes the farmers in the EU-Mediterranean Member States to feel more vulnerable. At the same time bilateral liberalisation of the agricultural trade within the Euro-Med Agreements although wished by the northern EU Member States and the MPCs, as it is promising for better market access, is feared by the farmers in the southern EU Member States due to the high similarity in their exports with the MPCs. Multilateral liberalisation is connected to the issue of preference erosion and thus seen with scepticism from the MPCs.

Due to this complexity and uncertainty on Mediterranean agricultural trade the need for deeper analysis emerges even more demanding.

3 Theoretical analysis of agricultural trade policies

This chapter theoretically analyses the agricultural policy reforms discussed in the previous chapter. It examines the effects of the latest CAP reform, the forthcoming free trade area around the Mediterranean basin and multilateral liberalisation involving the EU and the MPCs. Particular attention is paid to allocative and distributional effects. To keep the analysis less cumbersome the effects on single markets and several regions are addressed first. Effects due to market linkages, both vertical and horizontal, are discussed separately in the last section of this chapter.

To measure these effects the theory of applied welfare economics, and particularly the surplus concept, is employed. Social welfare is depicted as the sum of rent for producers, the rent for the consumers and the effects on the state's budget. The expenditure and/or revenue of the state influence the taxpayers' payments and thus overall social welfare.

The rent of producers is approximated using the notion of profit. In an output market the producer surplus is defined as the area above the supply curve and below the price line of the corresponding producing entity (JUST et al., 2004, p. 55; KIRSCHKE and SCHMITZ, 1990, p. 328). The producer surplus derived from an output can be alternatively measured in the input market in the same way that the rent of the consumers' is defined, assuming that the producers of any given output are the sole consumers of the input.

The consumer surplus is defined "*as the area under the demand curve and above the price line*" (JUST et al., 2004, p. 100). The notion of consumers' rent is more difficult to define since it is approximated by a monetary measurement of their willingness to pay for a certain commodity. Problems associated with simultaneous price changes and income changes are known as path-dependence problems and are dependent on which change is considered first (price or income change) – although this can be resolved in empirical research by using both paths to measure consumer surplus. A further remaining problem is the non-uniqueness of the money measurement of any change in welfare. Developments in the theory of applied welfare economics use the measurement of willingness to pay as an estimation of consumers' welfare and are based in the concept of the compensating and equivalent variation which are used as an alternative, approximation of willingness to pay.

It should be noted that in the following analysis considerations of the concept of willingness to pay are not used to measure welfare since they do not allow discussion of distributive and allocative effects. According to this concept social welfare consists of three components, the utility of consumers, the variable costs for production of the commodity that reduce the welfare of producers and the foreign currency refund or expenditure, which increases or decreases welfare respectively (HENRICHSMEYER und WITZKE, 1994, pp. 147-150).

3.1 CAP Reform: Allocative and distributional effects

The latest CAP Reform, as explained in the second chapter, can be summarised through three key words decoupling, modulation and cross compliance. It is difficult to quantify modulation and cross-compliance since quantitative modelling tools are insufficiently precise to fully measure their effects. Qualitative analysis is better suited for providing insights into the effects of cross-compliance. Hence this chapter focuses on the theoretical implications of decoupling direct payments and compares the impacts of coupled and decoupled direct payments.

To begin with, the term decoupling needs to be clarified. In the literature terms that have been used to characterise the direct payments are "fully decoupled" and "effectively fully decoupled". CAHILL (1997) characterises a policy as fully decoupled when it *"does not influence production decisions of farmers receiving payments and that permits free market determination of prices (facing all farmers, whether or not they receive income support)"*. LOPEZ (2001) argues that the concept of full decoupling is very strict and is only fulfilled when there is no change or distortion due to the application of the policy in the decision making processes of producers and consumers. A less restrictive concept is that of *"effectively fully decoupled"* which according to CAHILL (1997) refers to policies *"which result in production levels which are the same as those which would occur in the absence of support"*. LOPEZ (2001) comments that this concept focuses on equilibrium quantities, meaning that this policy would have no or almost no effect on trade and production. This implies that the decision making process of farmers can be affected by *"effectively fully decoupled payments"* but in such a way that no increase in production will take place.

As a basis for the discussion the development of the EU's policy package needs to be reviewed. The existing coupled payments were originally granted as compensation payments for price cuts under the MacSharry reform from 1992 onwards. This can be seen as a shift of subsidies from the actually produced quantity of a product to the actual product itself: for example, it is not the amount of tobacco produced per hectare that determines the level of benefits for the farmer, but the cultivation of tobacco. As a next step, the decoupling of direct payments – particularly in association with cross-compliance – shifts payments from single products to all agricultural production. Now it is not the planting of tobacco that qualifies for a payment, but the cultivation of agricultural land. The market impacts of public payments to farmers are determined by the production effects of the payments. This applies to every mode of granting payments, i.e. to both coupled and to decoupled payments. Hence, the theoretical challenge for analysing decoupling is to define a pragmatic term of *"production-effectiveness"* of direct payments, so as to work with and analyse the production-effectiveness of coupled and decoupled payments. This option seems more readily operationalised within an empirical model and has pragmatic benefits when discussing the implementation of the Single Farm Payment (SFP) scheme.

Theoretical analysis of agricultural trade policies 35

To elaborate more on the implementation of the SFP the focus is laid upon the decision processes of farmers over what to produce. A farmer decides for this product (or mix of products), which will maximise his profit. Coupled payments for single products per hectare or per animal can affect a farmer's relative competitiveness and can influence production, only if the possible profit of a product (or mix of products) including direct payments is the highest. In other words, supposing the cultivation of two different commodities (or two different product mixes) gives the same profit to the farmer, then he will decide what to produce based on the direct payments he expects to receive and he will choose for the product (or mix of products) for which the coupled direct payments are the highest. With some differences in detail, this also applies to decoupled payments: A farmer will produce (or will continue producing) if his agricultural profit exceeds alternative incomes. Acreage will be used for agriculture, if there is a possible product mix that gives a profit.

Therefore *production-effectiveness* can be defined as the share of direct payments which causes changes in the production structure compared to a situation with no direct payments. This can vary between 0% and 100%. The direct payments can be *ceteris paribus* converted into an increase in the producer's price that leads to the same changes of the produced quantity. Hence, a producer incentive price results from the farm gate price supplemented by the production-effective direct payment. Decoupled payments, which are only granted to active farmers, have a production-effectiveness greater than 0% if they increase farmers' profits or the profits of the acreage in such way that agricultural production is higher compared to a situation without coupled payments.

Building on these theoretical reflections, the next step is to analyse the impacts of coupled and decoupled payments on production at a market level.

Starting from a given farm gate price p_{FG} (Figure 3.1), a quantity q_1 is produced. A coupled direct payment scheme is implemented and it amounts to price effects equal to the distance of (p_4-p_{FG}) when converted into payment per unit. Assuming, that there is no production-effectiveness of this payment, the producer incentive price equals the farm gate price p_{FG}, while the produced quantity will not change. In this case, the granted payment tallies with the area (a+b+c+d). In terms of welfare, the increase in government expenditure is equal to the increase in producer surplus and the social welfare does not change. A decoupled direct payment with no production-effectiveness amounting to a fixed value equivalent to area (a+b+c+d) yields the same results.

Assuming a low production-effectiveness of these coupled and decoupled payments, the producer incentive price rises to p_2 and the produced quantity to q_2. Assuming a high production-effectiveness, the producer incentive price rises to p_3 and the produced quantity to q_3. Fixing coupled payments at (p_4-p_{FG}) per unit and decoupled payments equivalent to area (a+b+c+d), which equals the areas (b+c+d+f+g+h+k) and (c+d+g+h+k+n+r+s), then the following welfare impacts will result, taking the case with no production-effectiveness as a reference point (Figure 3.1)

Figure 3.1: Impacts of different levels of production-effectiveness of direct payments on producer's welfare

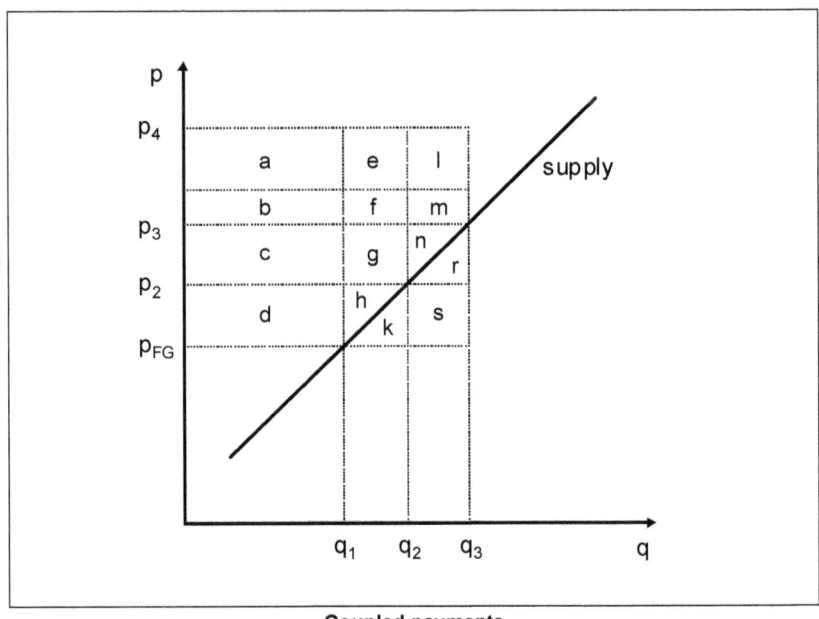

	Coupled payments	
	low production-effectiveness	high production-effectiveness
Δ producer surplus	= + (e+f+g+h)	= + (e+f+g+h+l+m+n)
Δ budget	= − (e+f+g+h+k)	= − (e+f+g+h+k+l+m+n+r+s)
Δ total welfare	= − k	= − (k+r+s)
	Decoupled payments	
	low production-effectiveness	high production-effectiveness
Δ producer surplus	= − a + (f+g+h)	= − (a+b) + (g+h+n)
Δ budget	= + a − (f+g+h+k)	= + (a+b) − (g+h+k+n+r+s)
Δ total welfare	= − k	= − (k+r+s)

Notes: Reference point for the welfare impacts is the case of no production-effectiveness
Source: KAVALLARI et al. (2005b)

Thus, with equal producer incentive prices, the loss of social welfare of coupled and decoupled payments is the same. However, with decoupled payments the producer surplus and government expenditure are lower. Welfare losses rise with a higher production-effectiveness. The maximum producer incentive price is reached at the p_3

level for decoupled payments, because the sum of direct payments is fixed. For coupled payments, p_4 is the maximum producer incentive price. Hence, the maximum impact of coupled payments on total agricultural production is higher compared to decoupled payments.

It should be noted that, based on the above, considerations of implementing the SFP in the EU and the effects on decision making processes are closer to the definition of "*effectively fully decoupled*" policies.

3.2 Free Trade Areas: allocative and distributional effects

The forthcoming Free Trade Area between the EU and the MPCs is a type of economic integration among two or more countries and is one form of Preferential Trade Agreement (PTA).

It is useful to more specifically define these terms. A Preferential Trading Agreement is a general term used to describe a broader spectrum of economic integration among two or more countries. Usually these agreements have a regional scope and are thus often called Regional Trade Agreements (RTAs). PANAGARIYA (2000) defines a PTA as a union where imports from countries that are signatories of an agreement are permitted at lower rates than imposed on imports from third countries. BURFISHER et al. (2003) and KRUEGER (1997) add that PTAs can have a partial scope and therefore focus on selected commodities, or they can have a total scope, meaning that the duty reduction covers all commodities produced by the parties to the agreement. An RTA can be agreed on a reciprocal basis, as for example the agreements that the EU has up to now with the MPCs, or can have a hub-and-spoke character, as for example the Everything But Arms (EBA) initiative.

A Free Trade Area (FTA) is a preferential arrangement in which the member states liberalise internal trade by setting all tariff rates between them to zero but maintain independence in setting external tariff rates (REED, 2001; PANAGARIYA, 2000). FTAs usually include detailed rules-of-origin specification. This prevents goods that enter the union from the member country that applies the lower external tariff rate being trans-shipped at a zero tariff rate to other members of the FTA. Instead a good is allowed to be traded duty free among members of the FTA as long as there is proof that it has been produced within the union. It has been observed that rules of origin are often very complex and can act as a trade barrier by specifying domestic thresholds on a commodity basis (BURFISHER et al., 2003). In the case of the Euro-Med Agreements the rules of origin are supported by the establishment of the Pan-Euro-Mediterranean protocol on rules of origin.

Further forms of economic integration are a Customs Union, a Common Market and an Economic Union. A Customs Union is an FTA, where all members impose the same external tariff on third countries (REED, 2001; PANAGARIYA, 2000). The next step of economic integration is accomplished through a Common Market, which in addition to being a customs union, also allows free movement of labour and capital within the union, as was the case with the European Economic Community in the early 90's. Harmonisation of the economic and fiscal policies, adoption of compatible monetary policies and common currency leads to an Economic Union, such as the EU.

PTAs have spread widely in recent years, although they violate the GATT/WTO most-favoured nation rule. By July 2007, a total of 380 regional PTAs had been notified to the WTO, with FTAs and partial scope PTAs accounting for 90 % and customs unions for nearly 10 % (WTO, 2007). It is worth noting that PTAs are only allowed by the WTO if they intend to promote trade between developing countries, as for example the Southern Cone Common Market (MERCOSUR) or if they fulfil the criteria of Article 24 of the GATT agreement, i.e. all trade is included and „*the general incidence of duties and other regulation of commerce*" is not higher or more restrictive against third countries than before the formation of the PTA (BURFISHER et al., 2003). From the above it becomes obvious that only FTAs or CUs are acceptable and partial PTAs are allowed only when they are concluded among developing countries.

Theoretical assessments of PTAs focus on trade creation, trade diversion and terms-of-trade effects due to the formation of an FTA or Customs Union. Trade creation effects refer to changes in the commodity trade within the PTA and take place when the domestic production of a member country falls and is displaced by imports from other member countries that produce at lower cost (REED, 2001; KRUEGER, 1997). Trade diversion effects are observed when third countries produce at a lower cost and when a member country of the PTA replaces imports from higher-cost partner countries with imports from third countries with lower costs. Finally terms-of-trade effects are due to changes in international prices caused by the formation of the PTA. Empirically it is shown that trade creation and terms-of-trade effects are welfare enhancing, while trade diversion effects reduce social welfare. Theoretical assessments of PTAs such as those by BURFISHER et al. (2003), PANAGARIYA (2000), KRUEGER (1997) and BHAGWATI and PANAGARIYA (1996) conclude that it is not clear whether PTAs are welfare enhancing or not and the welfare effects depend on the level of economic integration (i.e. Free Trade Area versus Customs Union), on the elasticities and on the initial reference scenario before the agreement came into force.

The welfare effects that occur due to a formation of a FTA are illustrated in Figure 3.2 (in page 41) which is a partial market diagram of three countries and one market [7]. In order to keep the diagram less cumbersome, it is assumed that the world consists of three countries, A, B and C, with A and B being potential FTA partners. It is assumed that firms in A, B and C supply at constant prices. Under perfect competition these prices represent the marginal and the constant average cost of production. Countries

---

[7] Throughout the analysis a price transmission elasticity of 1 has been assumed.

B and C are supposed to produce at the same cost, while A is the least efficient supplier. Country A is supposed to be a net importer, while countries B and C are net exporters, both exporting the same quantity to A. The domestic price in all three countries is the free trade price P_f.

In the reference situation it is assumed that Country A imposes a non-discriminatory tariff t per unit rate for all imports of the examined product thus setting the domestic price at the level P_i. The world market price P_w^1 which applies in both B and C is formed according to market equilibrium so that, imports from A (Q_2-Q_1) = exports of B (Q_4-Q_3) + exports of C (Q_6-Q_5) and exports of B=exports of C.

The next step assumes that A and B form a Free Trade Area. As a result country A abolishes the import tariff against country B, but maintains it for imports from country C. Further it is supposed that the FTA is governed by rules of origin which ensure that a lower tariff member cannot re-export duty free produce from third countries to the higher tariff member. Going beyond the framework of Viner and Meade, PANAGARIYA (2000) distinguishes three cases based on the total supply in A and B in relation to the demand in A after the formation of the FTA.

In the first case he assumes that the combined supply of a certain commodity in A and B is less than the demand in A. This implies that after the formation of the FTA, country B has an incentive to export more to country A (i.e. to export the maximum it can), since because the exports of B are not enough to satisfy the demand of A, the remainder of A's demand is satisfied by imports from C. A new equilibrium arises and thus the price in the free trade area is formed at the level P_i^1, while in country C a new world market price applies, P_w^2, which is lower than P_w^1 (dashed line in Figure 3.2). The demand for imports in A rises to Q_8-Q_7, the exports of B grow to $Q_{10}-Q_9$ while the exports of C are reduced to $Q_{12}-Q_{11}$.

These new prices benefit the consumers of country A since consumer surplus increases by the area a+b+c+d, while the producer surplus decreases by the area a. In country B the consumers suffer due to the increase of the domestic price and the consumer surplus decreases by the area n+o+q+r, while the producers benefit and the producer surplus increases by the area n+o+p+q+r+s. In country C the effects benefit the consumers with the consumer surplus increasing by the area u+v while the producer surplus decreases by the area u+v+w.

There are also changes in the tariff revenues. Country A used to enjoy a tariff revenue equal to the area c+f+i before the formation of the FTA. After the creation of the FTA, a quantity equal to $Q_{10}-Q_9$ is imported duty free, while for imports equal to $Q_{12}-Q_{11}$ country A collects tariff, the revenue from which is given as $(P_i^1-P_w^2)*(Q_{12}-Q_{11})$. The duty free imports are equal to the grey area illustrated in Figure 3.2, i.e. equal to o+p+r+s=e+f+h+i+k+l, while the tariff revenue of country A is equal to the area g+j+m. The taxpayers' effect in country A is thus equal to –c+g+j+m, while in countries B and C there are no changes in the effects on taxpayers since it is assumed that these countries have no policy for supporting exports (such as export subsidies).

In total the change of the social welfare is positive for country A, being equal to +b+d+g+j+m and attributable to trade creation effects. In county B the total welfare is again positive and equal to p+r due to trade creation effects. In country C the effects are negative and equal to –w, due to trade diversion effects. Thus it is not clear whether the global effects of the creation of the FTA are positive or not.

The second case that PANAGARIYA (2000) distinguishes is when the demand of A is equal to the supply of A and B. After the formation of the FTA country B has an incentive to export all of its supply to country A, country A imports only from its FTA partner, i.e. country B, while country C is crowded out from A's market.

In this case the trade creation and trade diversion effects are more profound for all three countries, as shown in Figure 3.3 (in page 42). The new price in the FTA is formed at the level P_i^2, which is slightly higher than the price P_i^1. The imported and exported quantities are Q_8-Q_7 and Q_{10}-Q_9 respectively and are equal to each other. While there are no changes in the direction of the effects regarding surplus of producers and consumers compared to the first case, the areas are now larger for country B and C and smaller for country A because of the higher domestic price level in the FTA countries, which results in a lower world market price (P_w^3) that applies for country C. This changes the taxpayers' effects in country A since all imports are duty free, country A no longer collects any tariff revenue, thereby negatively affecting its budget to the area c+f+i. Thus country A enjoys gains from trade creation but also suffers losses from trade diversion. Again it is not clear whether the entire world is better off or not from the FTA.

Finally the third case that PANAGARIYA (2000) examines is when the total supply of A and B exceeds the demand of A for the certain commodity. As in case two, country B now has an incentive to export the maximum to country A, crowding out country C. The welfare effects and the allocative and distributional effects are the same as in case two and are shown in Figure 3.3. Again it is not clear if the world is worse, or better, off as a result of the formation of the FTA.

The foregoing scenarios do not specify whether countries B and C trade with each other and whether they apply import tariffs or not. It could be assumed that country B for example also applies a non-discriminatory tariff t_b, which is lower than the tariff t, applied by country A. However as country B is assumed to be a net exporter, the pre-FTA price in this country is the world market price and thus nothing would change in the welfare effects shown above.

Theoretical analysis of agricultural trade policies

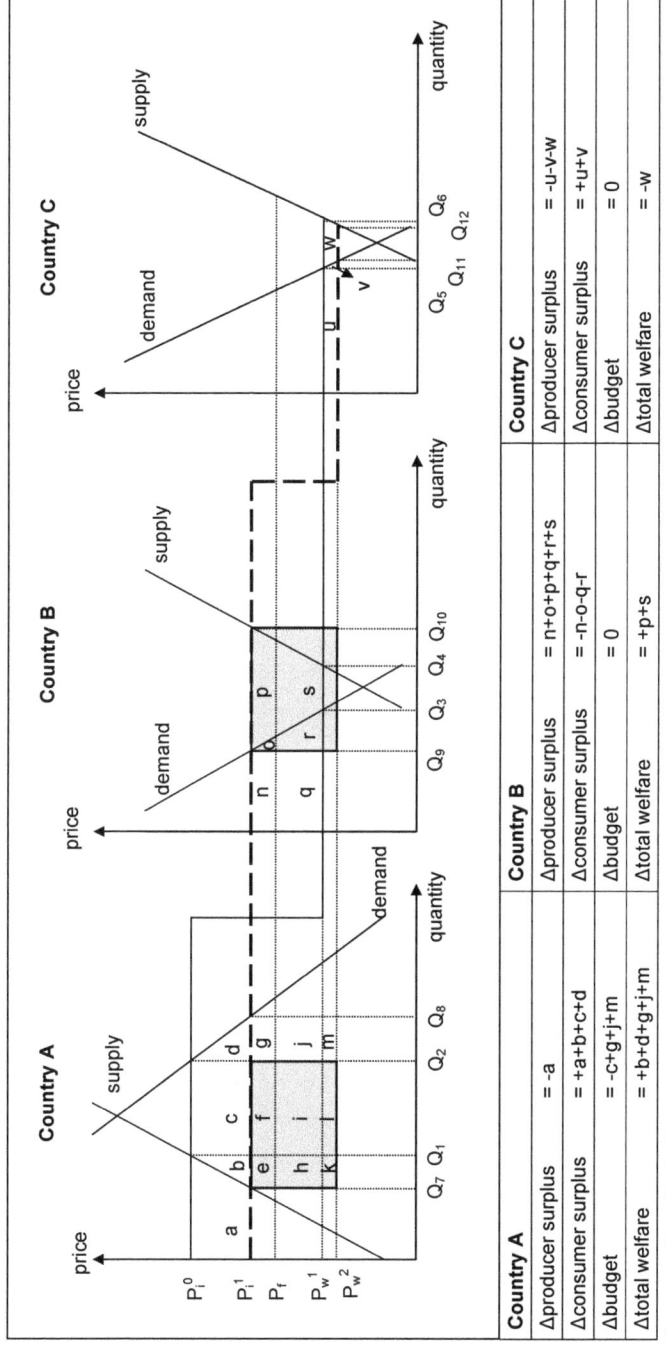

Figure 3.2: Effects of countries A and B forming an FTA, where countries B and C have the same production costs and the post-FTA demand of country A is met by imports from both B and C

Source: Own illustration

Figure 3.3: Effects of countries A and B forming an FTA, where countries B and C have the same production costs and the post-FTA demand of country A is met by imports from B alone

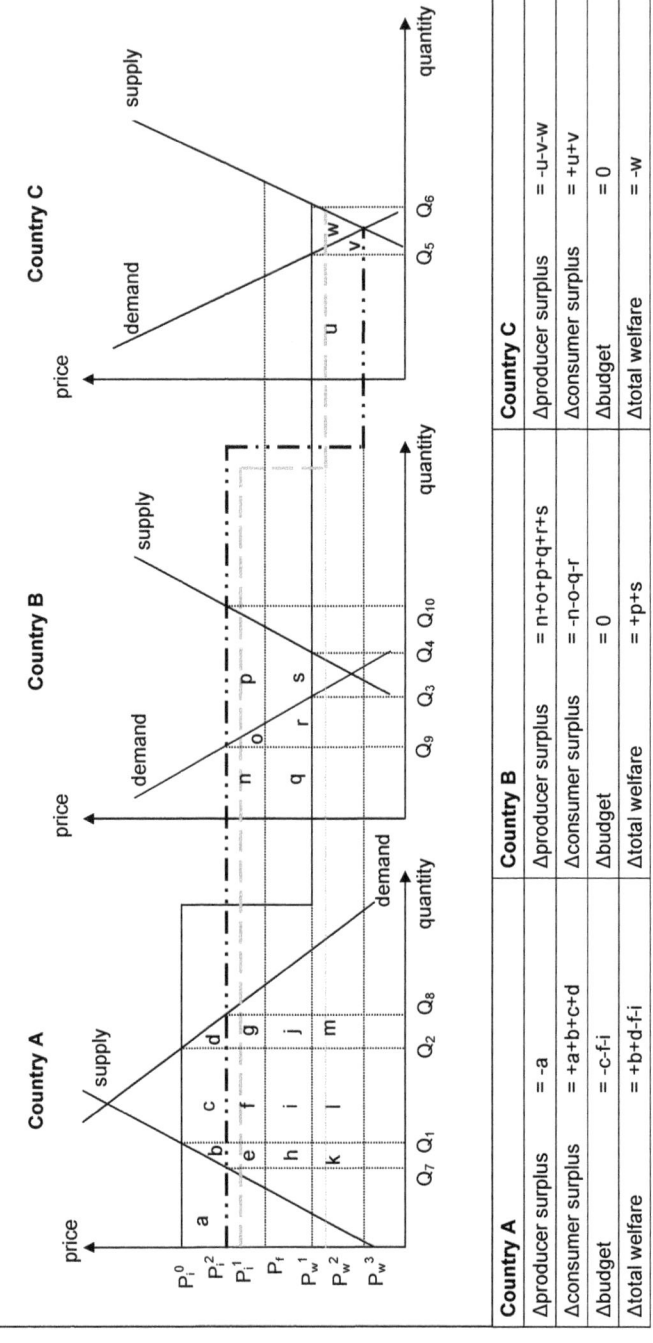

Country A		Country B		Country C	
Δproducer surplus	= -a	Δproducer surplus	= n+o+p+q+r+s	Δproducer surplus	= -u-v-w
Δconsumer surplus	= +a+b+c+d	Δconsumer surplus	= -n-o-q-r	Δconsumer surplus	= +u+v
Δbudget	= -c-f-i	Δbudget	= 0	Δbudget	= 0
Δtotal welfare	= +b+d-f-i	Δtotal welfare	= +p+s	Δtotal welfare	= -w

Source: Own illustration

Furthermore, the above analysis assumes that countries B and C have the same efficiency in producing a certain commodity. It could be in reality that one of them is more efficient than the other, for example country C could produce at a lower cost than country B. If this is the case then, in the pre-FTA situation, country A would import only from C, the most efficient supplier, but after the formation of the FTA with country B, country A would import from the higher-cost producers of country B, crowding out country C's producers. The resulting trade creation and, more so, the trade diversion effects are more profound for countries B and C than in the diagrams above and the entire world would be worse off. But if country A formed an FTA with country C instead of B, then both countries would enjoy the effects of trade liberalisation and the entire world would be better off.

Certain caveats still remain in this theoretical analysis of FTAs. It has been assumed that the level of protection against third countries applied by A (and B) remains unchanged after the formation of the FTA. Nevertheless this does not have to be the case and, combined with the ongoing WTO negotiations or with the common situation where one country is involved in more than one PTA, then the external tariffs change following the formation of an FTA. This will result in different allocative and distributional effects for the partner countries and for third countries. In addition, the average external tariff may vary due to the application of non-tariff barriers. Under an FTA, rules of origin often act as additional trade barriers (KRUEGER, 1997).

Another aspect that should be taken into account is that the formation of a PTA often provides market power to the new block of countries (BURFISHER et al., 2003) since it creates trade among its members, giving the partner countries within the block more market power than the individual members used to have. This can provide an incentive for small countries to enter FTAs, but also means that the new block can act as a price maker, influencing world market prices.

The above analysis shows that the welfare effects are not always profound for countries that form a PTA. Trade creation and trade diversion effects occur simultaneously and from the theoretical analysis it is not clear which of the two prevails. BHAGWATI (1999, pp. 14-15) concludes that the magnitude of the effects depends on the relative sizes of imports from each source combined with the expenditure on domestic goods and on the level of substitution between goods produced domestically and those produced by non-members.

Thus other motives must underpin the decision of countries to enter a RTA. CRAWFORD and FIORENTINO (2005) identify a variety of political, economic and security considerations. They argue that countries often use RTAs as a means of promoting deeper integration in terms of investment, competition, environmental and labour standards. The creation of a PTA is usually driven by a search to enter larger markets and this can be much easier at a regional or bilateral level than a multilateral one. Preferential access to markets can bring long-term advantages for suppliers, giving them an advantage over the competitors, especially with respect to trade in services. Discriminatory liberalisation can be particularly attractive for countries that fear international competition. It is accepted that developing countries might be willing to forego the benefits that GSP (Generalised System of Preferences) programmes promise them and instead enter into reciprocal PTAs with developed countries so as to access their markets and thus get a comparative advantage over their competitors. Moreover, membership of a PTA is thought to indirectly provide a certain security for foreign direct investment, which is of particular importance for countries with lower labour costs that enter into a PTA with a developed country. For example the EU established the MEDA programme under Council Regulation 2698/2000 to support investments in the MPCs (Official Journal of the European Communities, 2000). In this respect it seems that a PTA locks out international competition but locks in foreign direct investment. For smaller countries membership in a PTA seems to be a defensive necessity, while for larger ones it is essential so that they are not left behind. Finally political reasons (for example common defence programmes) can forge geopolitical alliances and cement PTAs.

These additional factors are not examined by the traditional theory of PTAs, illustrated by the partial market diagrams above. Nevertheless the New Trade Theory takes account of aspects connected to the formation of PTAs such as investment flows to partner countries, transfer of know-how, labour mobility and the liberalisation of production factors. In an empirical review of general equilibrium models applied to analyse the impacts of regional trade agreements ROBINSON and THIERFELDER (2002) conclude that the welfare effects are of a larger magnitude when models incorporate features of new trade theory and with these factors included aggregate trade creation is higher than trade diversion. These applied trade models often allow for differentiation of goods based on their origin, examine the issues of economies of scale and imperfect competition and take transportation costs into consideration.

It is clear throughout that a preferential supplier (country B in the above analysis) enjoys a preference margin as a result of the formation of the PTA. Yet it is worth exploring who in reality manages to capture the preference margin. GRETHE et al. (2005c) and FRANCOIS et al. (2006) argue that often in practice it is not the exporting partner country but the importing one, that gets the preferential margin, and this is dependent on the extent to which importers and their logistics sector exercise market power. The reduction of the preferential tariff due to the formation of an FTA can be used to sell at a lower price in the partner country in an attempt to expand exports. Alternatively, if the domestic price of the importing partner country is fixed, then it can be used to increase the received price for a certain quantity. This issue is of special interest when tariff rate quotas are applied or when a minimum import price system exists, as with the EU's entry price system for fruits and vegetables. These outcomes may well depend upon the way that licences are allocated to importers and exporters.

OLARREAGA and ÖZDEN (2005) show that exporters in developing countries received on average only one third of the tariff rent, and that this percentage is even lower in poorer and smaller countries.

The issue of the allocation of the preference margin is particularly relevant for the trade of fruits and vegetables between the EU and the MPCs, where the EU is the net receiver and the MPCs the net suppliers. GRETHE et al. (2005c) and GRETHE and TANGERMAN (1998a) argue that all the licences issued by the EU under the preferential arrangements have gone to European trading companies, suggesting that these companies accrue the preference margin. Nevertheless, because of the EU's entry price system for fruits and vegetables, monopolistic export structures often become established in the MPCs. This enables the MPCs' exporters to have a stronger negotiating power with the EU's trading companies, allowing them to attract at least some part of the preference margin. KALAITZIS et al. (2007) verify that most exports from the MPCs to the EU are organised through export-related institutions, such as associations of exporters or trading companies.

The MPCs are faced with welfare losses due to the formation of the Euro-Med Agreements as these will re-distribute the preferential margin to the exporters in cases where the EU is the net supplier, as is the case for cereals and livestock commodities.

The above analysis demonstrates that a country with a higher initial set of tariffs should expect welfare losses when it forms a PTA with a country with lower initial tariffs. This point is of particular relevance to the MPCs, which have higher initial tariffs and are joining a PTA with the EU, a low-tariff union that hegemonies trade. The MPCs can thus be expected to suffer from welfare losses when the Barcelona Agreement enters into full force. According to BHAGWATI and PANAGARIYA (1996), these loses will also depend on the volume of trade. They argue that higher volumes of trade lead to greater welfare losses for the partner countries. The depth of trade creation and trade diversion effects and their relative importance for the involved partner countries and the rest of world can only be specified empirically.

3.3 Multilateral liberalisation: allocative and distributional effects in a single market

The Mediterranean Partner Countries have shown strong interest in participating in the multilateral trading system that is currently under discussed. Although, at the time of writing, the negotiations of the Doha round have not come to a concrete conclusion, multilateral liberalisation is still on the agenda. Commitments to reduce or eliminate export subsidies and import tariffs and providing improved and non-discriminatory market access are currently being discussed and will eventually take place.

Free trade is considered to be the most durable and central economic policy and is recommended by economists. WONNACOTT and WONNACOTT (2005) suggest that an individual country should expect gains by abolishing the tariffs it applies to imports on the one hand and by getting better market access to other countries on the other. The second effect arises from reductions in tariffs applied by other countries on its exports. The following section draws out a theoretical explanation of both of these effects.

Figures 3.4 and 3.5 show the price, quantity and welfare effects of multilateral liberalisation on a single market. These effects will depend on the initial trade status of a country, i.e. whether it is a net importer or a net exporter. The reference system for the analysis is the currently existing agricultural policy. Barriers to free trade are imposed by various policy instruments, such as import tariffs, export subsidies, import quotas, entry prices or non-tariff barriers. The protection applied by large countries (price makers) suppresses current world market prices, keeping them artificially low.

In a large, net importing, country (Figure 3.4) the application of trade protection to domestic prices yields a price P_i, while because of the subsequent effects on the world markets, world market prices are lower than they would be under free trade conditions, at the level P_w^0. Multilateral liberalisation means that free trade conditions are applied to the markets, i.e. in the examined market the world price is arrived at through global demand and supply (or alternatively on excess demand and supply in a particular country or region) and would be higher than P_w^0, for example P_w^1. This price effect would lead this country to expand its imports from Q_2-Q_1 to Q_4-Q_3.

Consumers enjoy positive welfare changes, since they see their surplus being increased by the area a+b+c+d, while producers are worse off and face a decrease of their surplus that is equal to area a. The country no longer collects tariff revenue (equal to areas c+e), thus facing a reduction of its budget equal to this. Overall the deadweight losses of trade protection are recovered (i.e. triangles b and d) while the the effects on the terms of trade are negative and equal to area e. Thus, it is not clear whether or not the country will enjoy welfare benefits. This depends on the elasticities of demand and supply as well as world market prices, and specifically on the development of these before and after trade liberalisation.

Figure 3.4: Welfare effects of trade liberalisation on a net importing country

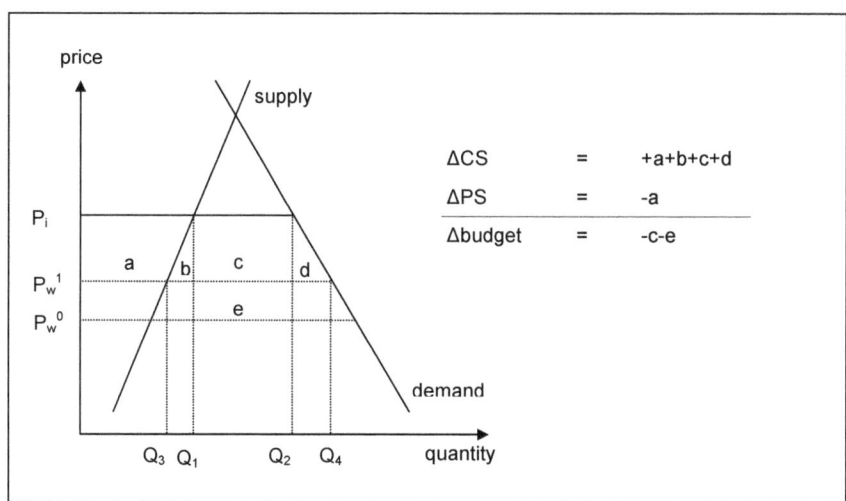

Notes: CS: consumer surplus; PS: producer surplus
Source: Own illustration

The results are slightly different for a large net exporting country, as illustrated in Figure 3.5. Due to its export subsidy, the country exports the quantity Q_2-Q_1 thus keeping the world market prices at the level P_W^0. Once trade restrictions are abolished and the export subsidy is eliminated, the markets function according to supply and demand, which results in decrease of the exported quantity to Q_4-Q_3 and an increase in the world market price to, for example, the level P_W^1.

The social welfare of a large, net exporting, country increases, not only because the deadweight losses of trade protection are recovered but also because the effects on the terms of trade are positive (shown by the area e+f+g+h+i in the Figure 3.5). The allocation of these effects among producers and consumers is similar to the large, net importing, country, although the magnitude is different. The consumer surplus rises, but not as much as under net importing conditions, while the producer surplus drops off more. The effects are shown by the areas a+b and -a-b-c for consumers and producers respectively. The state's budget is favoured by the free trade conditions as the burden of providing export subsidies is removed.

The increase of world market prices cannot lead to a change of the net trade status of a large country, as shown by Figures 3.4 and 3.5.

Figure 3.5: Welfare effects of trade liberalisation on a net exporting country

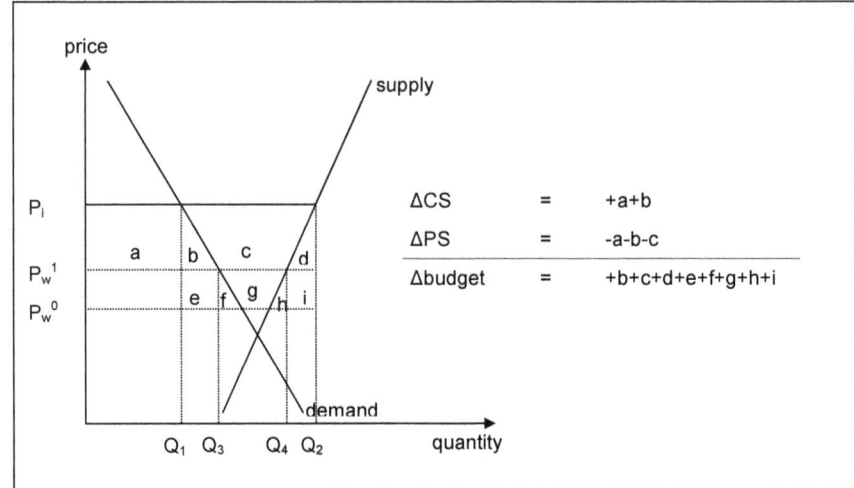

Notes: CS: consumer surplus; PS: producer surplus
Source: Own illustration

The effects are different for small countries, because they are price takers and cannot influence world market prices by applying a domestic policy. This is illustrated in Figure 3.6 where the world market price increases from P_W^0 to P_W^1 and these price effects lead to an upward and downward adjustment of the imported and exported quantities respectively.

The welfare effects on small countries depend on their net trade status, as shown below. For a small net exporting country the consumer surplus decreases by area f, while the producer surplus increases by the area f +g, resulting in overall welfare gains equal to the area g. In a net importing country the consumers will suffer a decrease of their surplus equal to the areas a+d+b, while the producers will gain a surplus equal to area a. As a result the overall welfare will decrease by area b. Finally when a change of the net trade status takes place due to an increase of world market prices, and the net importing country becomes a net exporting one, then the consumer surplus will decrease by the area c+e, the producer surplus will increase by the area c +d and the total overall change in welfare will be +d –e. Thus, it is not clear whether the welfare in this country will rise or not. It should be noted that, for the purpose of simplicity, it has been assumed that small countries do not apply trade policies which would result in different domestic prices but that their internal market price is the world market price.

It becomes clear that net importers gain from multilateral trade liberalisation while net exporters suffer from welfare losses. Within countries the allocation of these welfare effects is in favour of the producers. Consumers in countries that are net importers, net exporters, or experience a change of trade status suffer from losses, since they are faced with higher prices, while the producers enjoy an increase in their surplus because of higher prices.

Figure 3.6: **Welfare effects of agricultural trade liberalisation on small countries**

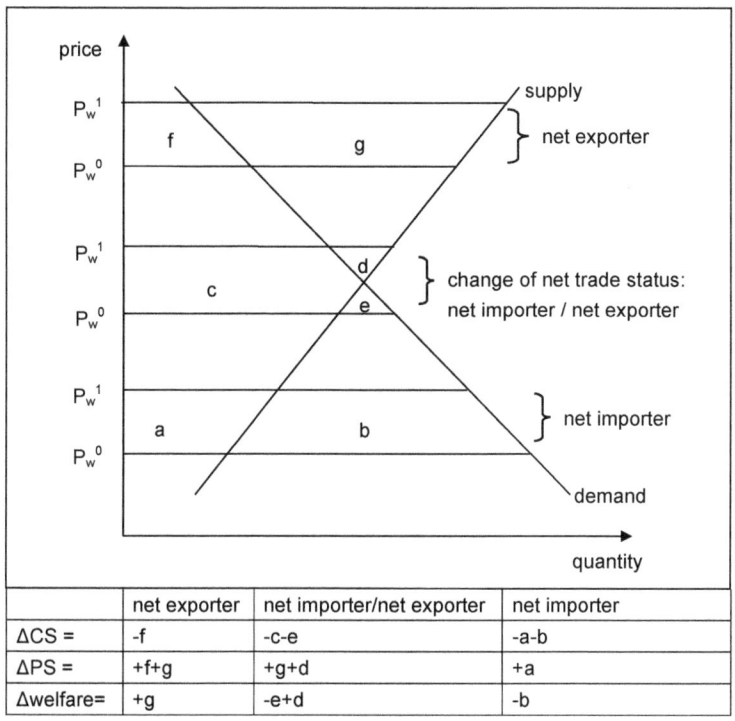

	net exporter	net importer/net exporter	net importer
ΔCS =	-f	-c-e	-a-b
ΔPS =	+f+g	+g+d	+a
Δwelfare=	+g	-e+d	-b

Notes: CS: consumer surplus; PS: producer surplus
Source: Own illustration based on PUSTOVIT, 2003

The assumption that these countries do not apply trade policies and are thus influenced solely by the increase of the world market prices does not always hold. Most such countries usually apply a certain degree of protection. Even more because many are involved in some form of preferential agreements (as discussed above), applying discriminatory protection against imports from certain countries, while providing preferential market access to some trade partners. In the case of the PTA between the EU and the MPCs, the latter already have a certain preferential access to EU markets even though the FTA is not yet fully activated.

In this situation the theoretical assessment of multilateral trade liberalisation becomes more problematic, with the effects depending not only on the initial net trade status of a country, but also on the type of the initial protection. The effects are connected to those of preference erosion and differ between partner and non-partner countries. FRANCOIS et al. (2006) explain that preference erosion effects arise from the reduction or elimination of tariffs on non-preferential suppliers.

To explore the allocation of the welfare effects that will occur in the event of preference erosion due to multilateral trade liberalisation, the same graphic is used as in Figure 3.2 (see Figure 3.7). It should be noted that the graphic of Figure 3.3 could be used again. The effects would be of the same direction, but of higher magnitude since Figure 3.3 illustrates an extreme situation, where the non-preferential partner is completely crowded out by the preferential one after the formation of the FTA.

For the welfare analysis the current status quo between the EU and the MPCs is used as the reference point, i.e. the PTA between countries A and B exists, resulting in price P_i^1 within the union for the partner countries and in a world market price of P_w^2 for the non-partner countries, because of the application by country A of a non-discriminatory tariff against all imports. As before, it is assumed that country A is a net importer and countries B and C are net exporters. Countries B and C have the same efficiency in producing the examined commodity, and country A is the least efficient supplier. This assumption (ceteris paribus) implies that under free trade conditions country A would import the same quantity from both countries, implying that after full multilateral liberalisation countries B and C will export the same quantity to A. The formation of the PTA allows country B to enjoy preferential market access to country A. As explained in the previous section, country A imports a total quantity Q_8-Q_7 due to the PTA, with a quantity Q_{10}-Q_9 being imported duty free from preferential partner B and a quantity equal to Q_{12}-Q_{11} from the third country C, on which a tariff revenue equal to the area g+j+m is collected. The price within the union is formed at the level P_i^1 and the world market price, applied only to country C, is at the level P_w^1.

Multilateral liberalisation means that free trade conditions apply in the market and thus the net importing country has no reason to prefer a specific country and import more from this one. In this example it imports the same quantity from countries B and C. The traded quantities are adjusted according to market equilibrium. Country A will need to import the quantity Q_{14}-Q_{13}, while country B sees a drop in its exports and will export quantity Q_{16}-Q_{15} to country A. Country C on the other side sees its exports increase to Q_{18}-Q_{17}. The new price that is formed and applies in all three countries is P_f.

The producers in the partner countries are worse off from multilateral liberalisation as they are faced with a decrease of the price they used to enjoy. In country A their surplus is reduced by area α and in country B by the area n+o+p. By contrast the producers in third countries, such as country C, benefit from free trade because the new

Theoretical analysis of agricultural trade policies 51

world market prices are at a higher level. Their surplus increases by the area +δ+ε+ζ+u+η+w.

The results on consumers are in the opposite direction. Consumers in partner countries A and B are better off, due to the decrease in price and benefit from an increase of their surplus by the area α+β+e+f+g+γ and n+o respectively. In third countries the effects on consumers are also negative and they are faced with a decrease of their surplus by the area -δ-ε-u-η.

Effects on taxpayers and the state's budget are only relevant for country A, since only this country that used to collect an import tariff. Liberalisation means that country A has to abolish its import tariffs and can no longer collect the tariff rent. The effect on its budget is negative and is equal to the decrease of the tariff rent (-g-j-m). To remind the reader, it is not relevant to this analysis whether country B applies an import tariff, or not or how high it is, because country B is a net exporter.

The overall change in social welfare in country A is shown by the area β+γ+e+f-j-m. Despite the positive welfare effects of trade liberalisation, country A losses the tariff revenue and any conclusion about whether the country will benefit from free trade or not depends on whether the loss exceeds the other benefits or not. The preferential supplier (country B) is faced with a partial loss of the original benefits under the preferential scheme, although the loss is less than the benefits of the formation of the FTA. FRANCOIS et al. (2006) explain that the loss is partial and not complete because the preferences granted to country B include benefits that are relative to the original tariff-driven equilibrium from a non-discriminatory tariff reduction by the importer. The third countries recover from the losses imposed by the preferential scheme and due to preference erosion will enjoy welfare gains of the area ζ+w.

Empirical analysis of the multilateral opening of markets is relevant for MPCs, not only because of preference erosion effects, but also as possible future strategy vis a vis the Euro-Med Agreements coming into full force. Empirical evidence from ROBINSON and THIERFELDER (2002) suggests that global trade liberalisation increases welfare more than forming an RTA. If this is also the case for MPCs, then they should focus on multilateral trade liberalisation rather than on maintaining and extending their preferences within the Euro-Med Agreements.

The effects would be more profound if countries B and C were not producing at the same costs and country C was more efficient than the preferential supplier (country B). In this the case, the welfare effects for the net importing and net exporting countries would be as shown in Figures 3.4 to 3.6, while for the preferential supplier the effects would be only partial and not complete.

Figure 3.7: Preference erosion effects of multilateral agricultural liberalisation in the two partner countries A and B and in the third country C

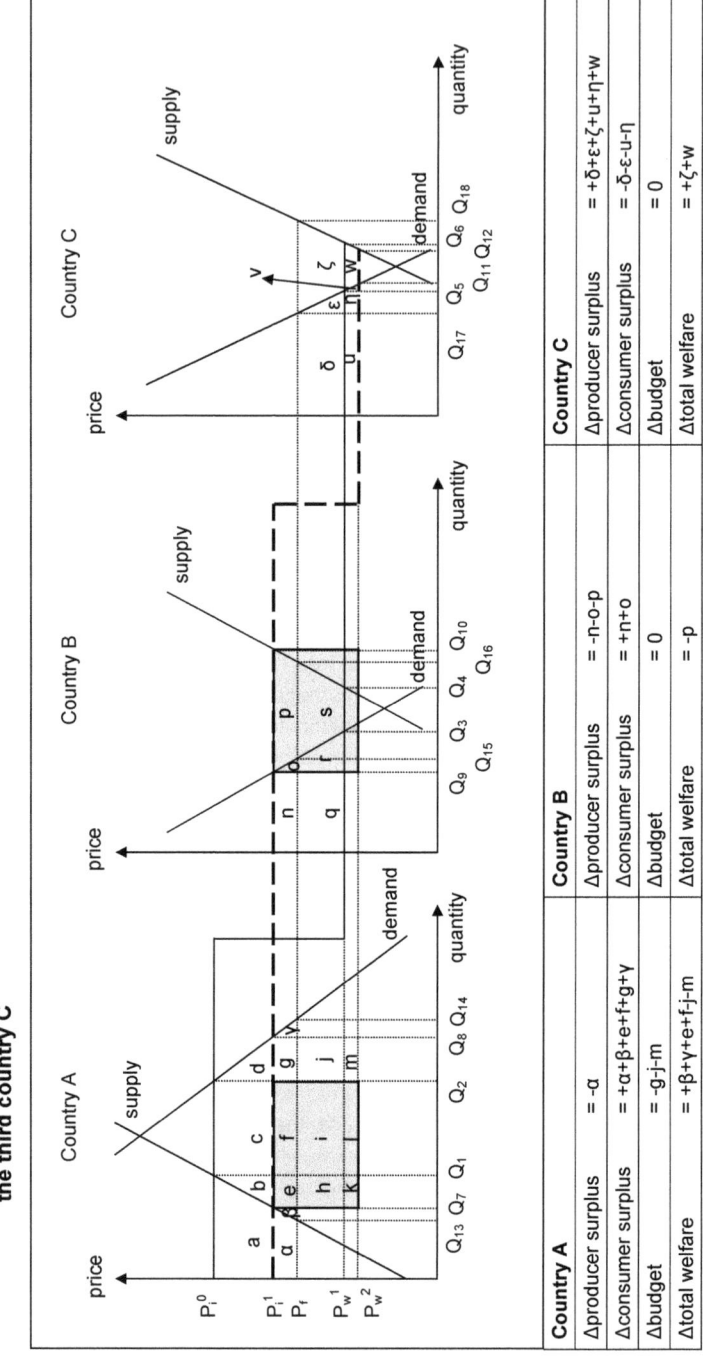

Source: Own illustration

From the above analysis it becomes clear that preference erosion is certainly beneficial for third countries and detrimental for preferential suppliers, especially if they are less efficient than third countries. The effects on the preferential importer are not clearly positive or negative. One limitation of this static partial analysis is that dynamic trade-related effects, such as further trade specialisation are not been considered. Countries in the G-20 group at the current WTO negotiations argue that preferences do not always favour small countries, because they encourage them to depend on a reduced number of uncompetitive products, discourage diversification and prevent the emergence of other suppliers (GARCIA ÁLVAREZ-COQUE, 2006a). If this is the case, then the elimination of preferences could be beneficial for small developing countries which supply the major markets in developed countries. This argument could be relevant for the MPCs, which at the moment are preferential suppliers to EU markets.

3.4 Market linkages

So far the theoretical considerations have only referred to single markets in several regions. However, agricultural trade policies applied to one particular agricultural market influence other agricultural markets. For example it is often observed that price changes related to a particular commodity lead to shift in supply or demand in a related market, which in turn results in the formation of a new equilibrium in the first market. Thus to complete the welfare analysis of agricultural trade policy instruments, it is necessary to discuss the effects caused by market interactions.

There are two types of market linkages that need to be taken into account. Markets can be linked vertically with an input – output relationship, or they can be linked horizontally if two products (two inputs or two outputs) are substitutable or complement each other in production and/or consumption. An example of vertically linked agricultural markets is that of cereals used as feed in livestock production. An example of horizontally linked markets is when two commodities are substitutes, competing, for example, for the same land resources because their production takes place under similar conditions, such as maize and cotton. Commodity substitution in consumption includes various types of meat (beef, pork, poultry meat). Finally, a traditional example of complementary products in consumption is that of sugar and coffee. This section discusses both effects in theoretical terms.

Figure 3.8 shows how linkages operate in vertically linked markets. Coarse grain as animal feed, is taken as an example for the input and beef as an example for the output market.

Figure 3.8: Vertical linkages between two agricultural markets

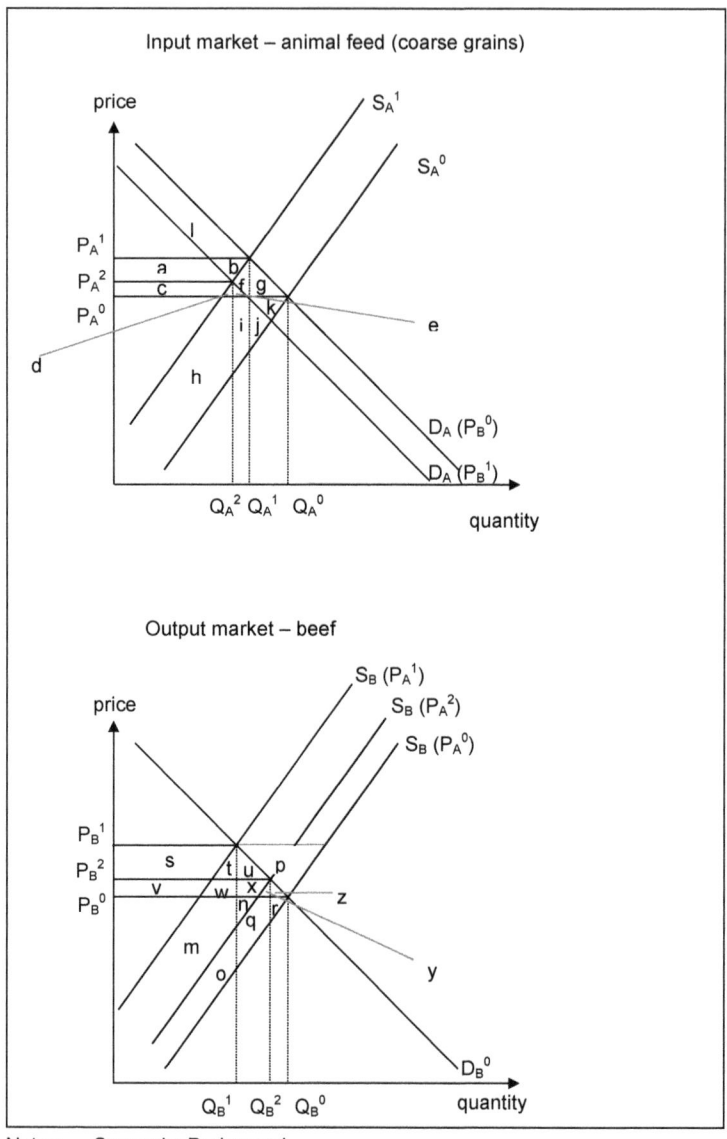

Notes: S: supply; D: demand
Source: Own compilation based on RONINGEN (1997), p. 239; JUST et al. (2004), pp. 312-322.

Initially quantity Q_A^0 of coarse grain is supplied at price P_A^0 following the S_A^0 supply schedule, while the supply schedule of beef follows the supply curve S_B (P_B^0). The demand schedule for beef is shown by the demand curve D_B^0. The demand and supply quantity of beef is formed according to the market equilibrium and is at the level Q_B^0 at price P_B^0. The demand schedule of animal feed is formed according to the demand for coarse grain from beef producers and is initially shown by the demand curve D_A (P_B^0).

In the beginning an exogenous shock in the input market leads to a shift of the supply curve for coarse grain leftwards from S_A^0 to S_A^1. This exogenous shock is due to a decrease in the supply of animal feed, which could for example be the result of crop failure or because coarse grains are used for alternate purposes (for example producing bio-ethanol). If there are price incentives then it could appear more attractive to coarse grain producers to sell their commodities for bio-ethanol rather than as animal feed. The new market equilibrium leads to a increase in the input price to P_A^1 and, because input becomes more expensive, the supply of beef meat upwards to S_B (P_A^1). A new output equilibrium is formed in which the consumed and supplied quantity of beef are reduced from Q_B^0 to Q_B^1 and the price of beef increases from P_B^0 to P_B^1.

The decrease in the quantity produced implies less demand for the input and thus on a second round the derived demand for coarse grain shifts leftwards to D_A (P_B^1). This sets a new equilibrium in the input market, with the input price decreasing from P_A^1 to P_A^2. The decrease in the input price makes the input cheaper for the output suppliers and leads to a third round where they shift their supply schedule downwards to S_B (P_A^2).

Consequently third, fourth and higher round effects would occur until the establishment of a final equilibrium in the input and output market.

To find the welfare effects for an entire sector, one needs to add all the rent changes throughout the markets (JUST et al., 2004, p. 321). Therefore, welfare calculations need to consider the surplus of the coarse grain producers, the surplus of beef consumers and producers (which is equal to the surplus of the consumers of animal feed, assuming that the entire production of animal feed goes into beef production).

The rent of the beef producers can be measured either in the output market or in the input market associated with derived demand. Even more the quasi-rent of the producers of the final goods can be measured in both input and output markets when the price changes are considered sequentially. Alternatively this rent can be measured in an intermediate market by constructing the equilibrium demand and supply curves, as shown by JUST et al. (2004), p. 315-318[8].

---

[8] The equilibrium supply curve is defined as „a supply curve that allows for equilibrium adjustment of input use and input price as the output price changes" (JUST et al., 2004, p. 315). In a similar way

The welfare effects are shown below, for just the first round effects of the above example (i.e. comparison of the equilibria $Q_A^0 P_A^0$ and $Q_B^0 P_B^0$ to $Q_A^1 P_A^1$ and $Q_B^1 P_B^1$ of the input and output markets respectively).

The surplus of the coarse grain producers decreases by the area h+i+j+k and increases by the area a+b+c. The surplus of beef consumers decreases by the area s+t+u+v+w+x+y+z.

The rent of the input consumers – output producers, measured in the input market by simultaneous price changes, decreases by the area a+b+c+d+e+f+g+l. This rent measured in the output market by simultaneous price changes decreases by the area m+n+o+q+r and increases by s+v. To measure these rent changes sequentially on both the input and the output markets, two options need to be considered. The changes can either first take place in the input market and then in the output or vice versa, as shown below:

1. a) Increase in the input price by keeping the initial output price constant ($P_A^0 \rightarrow P_A^1$, P_B^0 fixed): ΔR = -a-b-c-d-e-f-g as measured in the input market;

   b) Increase of the output price by keeping the already increased input price constant ($P_B^0 \rightarrow P_B^1$, P_A^1 fixed): ΔR = s+v measured in the output market. Overall the rent of the output suppliers – input consumers increases by the area s+v and decreases by the area a+b+c+d+e+f+g.

2. a) Increase of the output price by keeping the initial input price constant ($P_B^0 \rightarrow P_B^1$, P_A^0 fixed): ΔR = s+t+u+v+w+x+p measured in the output market;

   b) Increase of the input price by keeping the already increased output price constant ($P_A^0 \rightarrow P_A^1$, P_B^1 fixed): ΔR = -a-b-c-d-e. Overall the rent of the output suppliers – input consumers increases by the area s+t+u+v+w+x+p and decreases by the area a+b+c+d+e.

The sequential measurement is used as a practical response to data availability, since it is easier in practice to observe the welfare changes that occur between clearly determined price changes instead of between shifts of the supply and demand curves.

Determining the net social changes derived from the initial external shock in the input market involves aggregating the net effects, described above, to the individual actors. Thus the effects to the input and output producers, as reflected in the input market, result in a decrease of welfare by the area a+b+c+d+e+f+g+h+i+j+k+l. The effects on

---

the equilibrium demand curve can be defined as a demand curve that allows for equilibrium adjustments of output use and output price as the input price changes. More details on how the two curves are constructed and how welfare can be measured in the intermediate market are given by JUST et al. (2004), pp. 315-318.

Theoretical analysis of agricultural trade policies 57

the input and output consumers, as reflected in the output market, yield a decrease of the area m+n+o+q+r+t+u+w+x+y+z.

The relationship between two substitute commodities is shown in Figure 3.9. The example illustrates shows the inter-relationships between the cotton and maize markets in the EU-27.

Currently the EU-27 is a net importer of maize and cotton (cotton lint) (EUROSTAT, various years). The market for cotton in the EU is liberalised, since no import tariffs or export subsidies are applied, while direct payments are made to farmers following a deficiency payments scheme. The EU applies zero duties for maize too and provides subsidies to maize farmers. An indication of the degree of the maize market protection can be taken from the magnitude of the producer support estimate (PSE), which in 2006 was in the order of 162 million Euro (OECD, PSE database, 2006).

If year 2006 is taken as reference, then in the market for maize, the adjustments due to the Luxembourg reform of the CAP had already taken place. The supply scheme for maize is depicted by the supply curve S maize and the demand scheme is shown by the demand curve D maize. The decoupled producer subsidy lies at the level $P_{M,d}^0$, while the world market price is at the level $P_{M,w}^0$.

The CAP reform of the cotton market first started to be implemented in 2006. Thus, in Figure 3.9 the supply curve Scotton depicts the pre-decoupling production scheme and the price $P_{C,d}^0$ captures the farm subsidies prior to the introduction of the Single Farm Payment. Initially the world cotton market price is assumed to lie at the level $P_{C,w}^0$.

Decoupling of farm subsidies in the cotton market, as explained in chapter 3.1, can be translated into a drop of the producer incentive price and, depending on the level of its effectiveness, on production. Its effects on the behaviour of producers can be depicted in a partial market diagram with lower producer subsidies. This is illustrated in Figure 3.9 3.9 which shows a drop of prices paid to cotton producers to $P_{C,d}^1$. The price changes on the cotton market result in a shift in the maize supply curve towards the right, since it becomes more attractive for farmers to replace cotton with maize. A pre-condition for such a shift is that the applicable maize prices (including producer subsidies) are higher than the reduced cotton prices.

Figure 3.9: Horizontal linkages between two agricultural markets

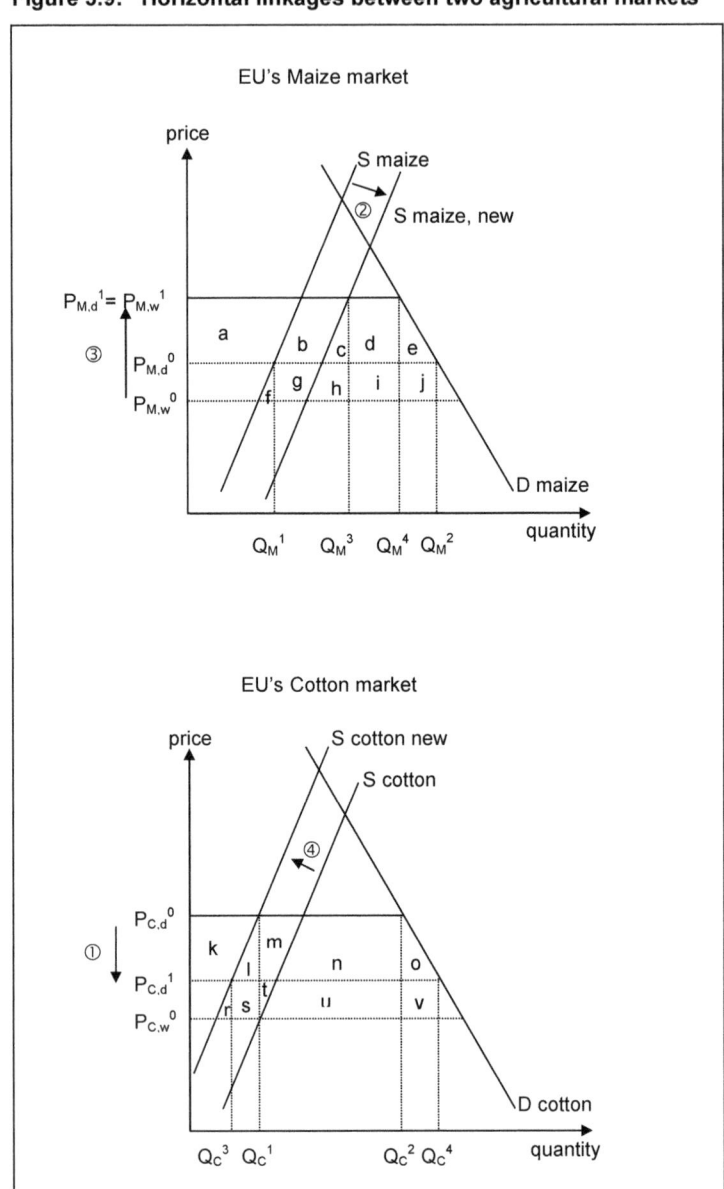

Notes: S: supply; D: demand
Source: Own presentation

Theoretical analysis of agricultural trade policies

In a second step, the world market price for maize increases, as we are currently observing. A possible explanation for this increase could be developments in the bioethanol market, which have increased the demand for maize, driving up world market prices. To keep the diagram less complicated, it is assumed that due to an exogenous shock the world market prices increase while the demand schedule remains unchanged. The increase in the world market maize price is so high that the new world market maize price is higher than the domestic price in the EU, so the producer subsidy is no longer relevant for maize producers. The new world market price $P_{M,w}^1$ is simultaneously the new domestic price $P_{M,d}^1$.

Due to price effects in the maize market a new round of effects takes place. In a following step EU producers would replace cotton with maize, as shown by the upward shift of the cotton supply curve from S cotton to S cotton new.

Obviously, subsequent changes will continue to take place until the markets are stabilised and until a new market equilibrium in both markets is established.

The price changes lead to adjustments in the supplied and demanded quantities of maize and cotton, giving rise to changes in the quantities traded. In the reference situation a quantity of $Q_M^2 - Q_M^1$ of maize was imported, while in the final situation the imported quantity is reduced to the level $Q_M^4 - Q_M^3$. In the cotton market the imports increase due to the sequential price changes and, after the shifts of the supply curve, the EU will import a quantity equal to $Q_C^4 - Q_C^3$ (as opposed to the original imported quantity $Q_C^2 - Q_C^1$).

The allocation of welfare is different in these two markets and can be simultaneously measured in each market. The overall social welfare for the sector can be derived from the aggregating the effects in each market.

In the maize market producers enjoy welfare gains equal to a+b+f+g since not only the domestic price increases, but the supply curve also shifts to the right. By contrast consumers suffer from the higher price and their surplus is reduced by the area a+b+c+d+e. Due to the increase in the world market price the effects on the taxpayers are positive. The state no longer needs to finance the producer subsidy and this leads to gains of the area g+h+i+j. The overall change of welfare is equal to -c-d-e+2g+h+i+j.

In the cotton market the impact of the different price adjustments is only felt by producers and taxpayers, since consumers enjoy the world market price, which remains stable throughout the sequential adjustments. The producer surplus decreases by the area k+l+m+r+s+t. The effects on taxpayers are due to the different expenditure on the producers' subsidy. Initially taxpayers were burdened with expenditure equal to m+n+o+t+u+v, while in the final stage their burden equals the area s+t+u+v. Therefore the change of the state's budget is m+n+o-s-v. Overall the welfare change is equal to –k-l-r-2s-t-v+n+o.

The sequential measurement of the overall welfare on both markets requires talking into consideration the welfare effects on producers and consumers of each price change that takes place, while the effects on taxpayers are measured individually for each market. Again the aggregation of the welfare effects of each step and on taxpayers gives the overall welfare effects. The price changes take place in two sequential steps:

1. Decrease of the domestic price for cotton farmers from $P_{C,d}^0$ to $P_{C,d}^1$, while the initial maize price remains unchanged. The effects on the producers are measured on the cotton market and are equal to a reduction of their surplus by the area k+l+m. There are no effects on consumers.

2. Increase of the maize price from $P_{M,d}^0$ to $P_{M,d}^1 = P_{M,w}^1$, while the, now lower, cotton price remains unchanged. The welfare effects are now measured in the maize market and for producers the change of surplus is equal to a+b+c, while for consumers it is equal to -a-b-c-d-e.

The effects on the taxpayers are as mentioned above. In the maize market the change of the state's budget is equal to g+h+i+j and in the cotton market equal to m+n+o-s-v.

Through the above analysis one can trace third and higher-round effects that can be considered as successive approximations of sequential market responses in related markets. To capture these effects empirically, TSAKOK (1990) suggests using cross-price elasticities in quantitative modelling techniques and claims that multi-market analysis can provide an appropriate framework for analysing simultaneous changes in linked markets. This type of approaches is used to analyse the properties of different market equilibria that come about as a result of exogenous changes. This approach assumes that other factors that can influence the production remain unchanged, (i.e. ignoring capital changes or technical progress) and in this respect it is a comparative static one.

The analysis in this chapter demonstrated that theoretical considerations are alone not enough to reveal the precise allocative and distributional effects on the Mediterranean countries (both in EU and non-EU countries) due to the already described agricultural policy reforms. How the trade flows are going to be formed, who will benefit and who will lose and how much the taxpayers are going to be burdened or not by the reform of the CAP, by the Barcelona Agreement and finally by multilateral liberalisation is without empirical analysis not easy to answer.

4 Empirical studies of agricultural trade policy reforms in the Mediterranean basin

This chapter reviews existing empirical assessments of trade and agricultural policy reforms that are relevant for the countries round the Mediterranean basin. After a presentation of the main modelling tools and their elements, there follows a literature review that focuses mainly on studies that have analysed the impact of changing trade policies. These are mostly focused on trade liberalisation, at the bilateral and multilateral level between Mediterranean countries, and only a few studies have analysed the latest CAP reform and its likely impacts on Mediterranean countries. The last section of this chapter presents the trade policy model AGRISIM that will be used for the empirical analysis within this study.

This chapter aims to identify the type of applied modelling tools that have been used and to explain how such policy reforms have been formulated within model scenarios. It also attempts to show existing gaps in the literature, and suggests the elements that a model should entail in order provide a satisfactory analysis of trade policy reforms as discussed theoretically in the previous chapter. It argues that the AGRISIM model incorporates those elements, can fill these gaps and provide a robust empirical analysis.

4.1 Basic elements of modelling tools

The impact of agricultural policies is assessed both *ex post* and *ex ante*. The *ex-post* studies describe the status quo and the historical development of trade and attempt to identify key factors that influence observed changes in trade flows. By contrast, the *ex-ante* studies try to answer "what if" questions and try to estimate likely changes through formulating plausible scenarios and giving insights on the impacts of expected policy changes.

The *ex post* studies are usually accomplished with the help of gravity models or other econometric techniques, such as regression analyses or the calculation of indicators, such as export similarity and trade concentration indices.

The gravity equation was first developed by TINBERGEN (1962) and PÖYHÖNEN (1963) and ever since has been applied by a number of authors to explain international trade flows due to migration, foreign direct investment or the existence of preferential trade agreements. It is based on the premise that a flow from an origin to a destination is explained by economic forces at the flow's origin and destination and economic forces that either aid or resist these flows (BERGSTRAND, 1985). NIELSEN (2003) explains that the main advantage of gravity models is that they are simple and intuitive but, due to their character the analysis is restricted to one-way trade. ANDERSON and WINCOOP (2003) note that due to their lack of theoretical foundation the results of gravity models are biased by omitted variables and that the estimated parameters cannot be used in comparative static exercises.

Ex ante analysis typically uses sector-specific or economy-wide models. While economy-wide models focus on the entire economy the sector-specific models depict effects in one particular economic sector. More specifically, the economy-wide models can be divided in three main categories: macro-econometric models, input-output models and general equilibrium models[9] (TONGEREN et al., 2001).

Macro-econometric models, as the name reveals, examine the macro-economic environment, for example, through the development of inflation or of exchange rates and do not focus on the agricultural, or any other specific sector.

Input-output models assess the impact of a change in the final demand of a given sector on all sectors of the economy (SADOULET and DE JAVRY, 1995 pp. 285-288). They are based on input-output tables, i.e. on a matrix that includes the activity and commodity accounts and the interactions between them. The technique of the input-output modelling was initially developed in the '30s by Vassily Leontief and is known as the Leontief model but since then it has been extended and modified so that the derived multipliers more accurately represent the multiplier effect on the economy (SADOULET and DE JAVRY, 1995 pp. 285-288). Despite these advances input-output models cannot capture welfare effects.

Equilibrium models, either partial or general, are more widely used in the analysis of trade policy reforms.

CGEs can be defined as models that capture the interactions between economic activities within a country, a region or the entire world by focusing on the intra and inter-regional interdependencies of markets and actors (BROCKMEIER, 1999). This allows for the examination of the allocation of resources which could arise, for example, from changes in international prices due to a reform of trade policy. They require an extended data base since they need to be fed with data for all economic sectors and for the transactions between them. Social Accounting Matrices (SAMs) or alternatively input-output tables are used as input data for CGEs. A SAM is a square matrix and differs from an input-output table in that the later table is a sub-set of a SAM, containing only the activity and commodity accounts (SADOULET and DE JANVRY, 1995 pp.285). When input-output tables are used as an information source for a CGE, then additional data on the behaviour of producers, consumers, importers, exporters and possibly other economic agents are required (TONGEREN et al., 2001). Due to the large amount of background data that CGEs require, global CGEs cannot be easily developed by individuals and in practice they are developed either by international

---

[9] Alternative names for the general equilibrium models are computable general equilibrium models (CGE) or applied general equilibrium models. Hereafter the term computable general equilibrium model will be used.

organisations or by consortiums that can afford to construct SAMs for specific regions. Some CGEs have been developed by individuals, but these have usually focused only on one country and are based on the SAM of this particularly country, thus do not provide an adequate regional picture. Equally the commodities within the economic sectors are often aggregated at a higher level, either due to data restrictions or just in an attempt to keep the models manageable and do not give detailed insights on a particular economic activity and economic sector. Therefore when discussing a particular economic sector, and even more so a limited number of specific commodities, it is more common to employ partial equilibrium models.

Partial equilibrium models (PEs) by definition focus on one part of the economy i.e. on one economic sector and do not take into account interrelationships and interdependencies between different economic sectors and the different inputs and outputs of the economy. Prices and quantities in the rest of the economic sectors and markets are not examined, and are assumed to be unaffected by policy change. The partial equilibrium models have the advantage of allowing a relatively transparent analysis of trade policy issues, since they focus on a very limited set of data on prices and policy variables (FRANCOIS and HALL, 1997 p. 122). Moreover they allow a higher disaggregated representation of the commodities within the economic sector studied than CGEs.

Equilibrium models (both partial and general) can be characterised as comparatively static or dynamic. The comparative-static approach involves comparing different equilibria that occur from a simulated policy shock, for example a change in an exogenous variable (FRANCOIS and REINERT, 1997 p. 6). When changes on investment and technology are captured, for example through shifts in the demand and supply curves and slope changes, then the approach is a dynamic one (TSAKOK, 1990 pp. 180-181).

Equilibrium models can be also characterised as multi-regional based on their regional focus and as multi or single commodity depending on their commodity composition.

The structure of the model can allow the assumption that the goods or commodities are homogenous, meaning that domestically produced and imported goods are considered as perfect substitutes in demand, or that they are imperfect substitutes (FRANCOIS and REINERT, 1997 p. 5). The latter assumption is known as Armington assumption (ARMINGTON, 1969) and is based on the use of constant elasticity of substitution functional forms to describe the preferences among imports from various countries.

Equilibrium models can also be divided into spatial and non-spatial ones. TONGEREN et al. (2001) explain that homogeneity is often connected to a pooled market in non-spatial models where it is known what each actor brings to, and takes from, the market. In the case when the products are differentiated according to their origin, then the price gap between the domestic and the foreign goods reflects the domestic tariff and the transport costs from one country to the other (or alternatively from one region to another). This modelling approach is known as a spatial one (BOUËT, 2006a).

Over the years several global multi-commodity equilibrium models have been developed and have been supported by international organisations. Examples of such models are the COSIMO model developed by the Food and Agricultural Organisation (FAO) of the United Nations (FAO, 2005), the AGLINK model, by the Organisation for Economic Co-operation and Development (OECD, various years; CONFORTI and LONDERO, 2001; TONGEREN et al., 2001[10]), the ATPSM model (Agricultural Trade Policy Simulation Model) by the United Nations Conference on Trade and Development (UNCTAD) (UNCTAD, 2004) and the model of the Food and Agricultural Policy Research Institute (FAPRI) (WESTHOFF and YOUNG, 2001). Further models have been developed by international networks of university-based scientists, such as the Global Trade Analysis Project (GTAP) (HERTEL, 1998) or within research projects like the models WATSIM (KUHN, 2003) and CAPRI (BRITZ, 2005) from the University of Bonn, ESIM from the University of Göttingen (BANSE et al., 2004) and CAPsi (LEDEBUR et al, 2005) from the Federal Agricultural Research Centre of Germany (FAL). Over the same period reviews and evaluations of these global models have increased as well, with the study of TONGEREN et al. (2001) being worthy of note.

Last but not least, *ex-ante* analyses have also been carried out with non-linear optimisation production models. These models belong to the family of mathematical programming models. BAUER and KASNAKOGLU (1990) and HECKELEI (2002) explain how these types of models determine input allocation to various production activities and allow for the modelling of technological and political constrains. They are programmed using Positive Mathematical Programming (PMP) techniques i.e. allowing calibration to observed behaviour within a base year.

All these modelling tools and methodological choices offer advantages and disadvantages in their application and should be seen as complementary approaches. In practise the choice of a model depends only on the problem that needs to be solved and on data availability and the available resources to develop it.

---

[10] An extended version of this study has been published as TONGEREN et al. (1999 and 2000).

4.2 Literature Review

The studies reviewed in this section are presented based on their main focus and on the methodology they have applied to assess the impact of the reformed agricultural trade policies empirically on the Mediterranean countries.

4.2.1 Empirical assessments of the latest CAP Reform relevant to Mediterranean countries

The latest CAP Reform has been topic of interest in recent years for agricultural economists, who have carried out various analyses using partial and general equilibrium models. A number of studies have employed static partial equilibrium models to assess the impacts of the latest CAP reform and particularly the introduction of Single Farm Payments. These include studies by LIPS (2004), KLARE and DOLL (2004), the EUROPEAN COMMISSION (2003); BINFIELD et al. (2003) and the FAA (2002). The general equilibrium studies mostly involve applications of the GTAP, as for example the studies by BROCKMEIER et al. (2003) and MEIJL et al. (2002). Wider views on modelling exercises of the Luxembourg Agreement are given by BALKHAUSEN et al. (2008)[11], while a number of papers review models applied for the analysis of the EU's common agricultural policy. These include a number of working papers from the Italian Institute of Agricultural Economics. ARFINI (2001) has evaluated positive mathematical programming models, CONFORTI (2001) main partial equilibrium models, DE MURO and SALVATICI (2001) multi-sectoral models and SCKOKAI (2001) econometric models.

The studies analyse the first wave of the CAP Reform, from the Luxembourg Agreement and discuss its impacts on typically northern agricultural commodity markets, such as cereals, rice and livestock commodities, such as milk, beef, pork and poultry meat. They make predictions for either the EU as one region (BROCKMEIER et al. 2003; EUROPEAN COMMISSION, 2003; FAA, 2002; LIPS, 2004), or specific countries like Ireland (BINFIELD et al., 2003 and BREEN et al. 2005) or the Netherlands (MEIJL et al., 2003).

The modelling of the decoupled direct payments that were introduced with the Single Farm Payment is a key methodological aspect that requires attention. In partial equilibrium models such in this of FAPRI, the decoupling is modelled through the application of a multiplier that functions in a similar way to the multiplier of production-effectiveness, described in chapter 3. A description of way in which the FAPRI model approaches modelling decoupling can be found in BINFIELD et al. (2005).

---

[11] An older version of this paper has been published as BALKHAUSEN et al. (2005).

Due to the regional composition of the models that have been employed to analyse the CAP Reform, the regional effects on the Mediterranean EU Member States on the one hand and on non-EU Mediterranean countries on the other are not revealed.

The second wave of the CAP reform (for cotton, olive oil and tobacco) that might affect the agricultural sector of the EU Mediterranean Member States and other Mediterranean countries has only attracted limited attention from agricultural economists. Three noteworthy studies examine the impacts of the CAP reform on cotton and tobacco and are discussed in more detail in the next three paragraphs.

The modelling exercise of KARAGIANNIS (2004) focuses on the effects on the EU's cotton producers of the CAP reform on cotton. The author developed a single market single country synthetic partial equilibrium model, which was applied to both Greece and Spain. In the model the policy variables, which are the entitlements per producer for the eligible areas are converted from a per-hectare to a per-tonne basis. The new area payments are treated as input subsidies that shift the supply curve of land downwards and result in a decrease of the rental price of land, which in turn increases the quantity of land used for cotton production. The model treats the new area payment as a subsidy targeted at one production factor and the market price support as a subsidy spread evenly across all inputs. The model uses the period 2000-2003 as its reference period and gives results on production and producers' income, which is estimated by the producer surplus. The application of the Single Farm Payment in both countries is expected to result in a lower supply, with the effects on the income of the cotton growers being positive, although to varying degrees, depending on the supply elasticity. With a supply elasticity of 0.35, the average change in the producer surplus is 3.6 % in Greece and 6.1 % in Spain, while with a supply elasticity of 0.7 the change is 12.1 % and 16.8 % in the two countries respectively.

STOFOROS and MERGOS (2004) have used a partial equilibrium model of the EU's tobacco sector to examine the impacts of the EU tobacco policy reform on the EU's producers, consumers and taxpayers. The authors have estimated changes on production, consumption and on the EU's self-sufficiency and net trade based on two future scenarios for 2010. The first scenario (baseline scenario) assumes the continuation of the EU's tobacco policy, while the second involves the total elimination of deficiency payments. The results of the second scenario (expressed as changes from the baseline scenario) indicate a 20 % decrease of production and an 8 % decrease in consumption. The authors explain that abolition of the tobacco deficiency payments will affect the EU's self-sufficiency, which will result in a rise in prices and thus a decline in consumption. Consequently, the imports of tobacco into the EU will increase and the EU's self-sufficiency will decrease from 0.45 in 2003 to 0.39 in 2010. The effects on producer and on consumer surplus are negative (losses of 226 and 88 million Euro respectively) but are positive for the entire EU budget (an increase of 267 million Euro). The demand, supply and price elasticities which underlie the modelling exercise were estimated by the authors.

A further study on the impacts of the CAP Reform on Mediterranean commodities has been carried out by ARFINI et al. (2005). Using a farm model based on Positive Mathematical Programming the authors examine the Italian tobacco sector and elaborate the effects of the reform of the tobacco CMO on harvested surfaces, farm incomes and overall employment for a sample of farms derived from the Italian FADN (Farm Accounting Data Network). Because the study does not employ a trade model, the effects on trade flows and on welfare cannot be derived and the results only focus on Italian farms that grow tobacco. The results indicate that decoupled aid would not justify a continuation of tobacco cultivation and expects a reduction in production of up to 95 % in the examined regions. This decline in production is expected to have negative results on employment, with non-family workers expected to be the most negatively affected group.

The EU Parliament commissioned GARCIA ÁLVAREZ-COQUE (2006b) to carry out an expert estimation of the likely effects of possible reforms to the support regime for fruits and their static impacts on the EU's fruit and vegetable markets. However, these reform proposals are only on a preliminary level and did not reflect the substance of the EU Commission's proposals for the reform of the fruit and vegetable CMO that was later published (Commission of the European Communities, 2007a and b).

To date there has been no examination of the effects of the Luxembourg Agreement through *ex-ante* modelling exercises, since the effects of this reform just begin to become evident (in economic year 2006/07) and thus the observations are not sufficient for such an analysis.

4.2.2 Empirical assessments of trade liberalisation that is relevant to Mediterranean countries

Studies on trade liberalisation around the Mediterranean mainly focus on analysing the Euro-Med Agreements. The impact of this has been assessed both *ex-post* and *ex-ante* as association agreements came into force at different times among the Mediterranean Partner Countries. Most of the ex-post studies refer to the impacts in the entire economy, and not particularly on the agricultural sector, since trade liberalisation has already taken place and is mostly related to the secondary and tertiary sectors.

Ex post studies

Econometric analyses

MUAZ et al. (2003) reviewed and evaluated the consequences of the Euro-Med Agreements on Jordan, Palestine, Syria, Lebanon and Egypt, with particular reference to the effects on the economic development of the agricultural sector. Through a PAM approach the authors sought to quantify the comparative advantage (or disadvantage) of eleven horticultural crops[12] in these countries and to quantify the main economic and social implications of expanding the export volumes of the most competitive crops, in terms of national income, employment, investment and the use of resources (especially water). The analysis shows that, due to the Euro-Med Agreements, the EU is the major destination market. The calculation of the Domestic Resource Cost (DRC) reveals the MPCs to have comparative advantages in horticultural production, mainly because of favourable climatic conditions, competitive labour costs and their proximity to the EU markets. At the same time, market analyses in Germany, France, the Netherlands and the UK, based on the concept of profitable demand, confirms that there are market windows for the MPCs, mainly during the winter months and in the beginning of spring and the end of autumn, i.e. in months were there is no production from the EU Mediterranean member states. For example, the estimated profitable demand in the Netherlands for green beans was for 24,041 metric tons between December and April, whereas in the UK, France and Germany the demand was for 12,736; 12,782 and 28,586 tons respectively. Meeting profitable demand for green beans, strawberries, melons and table grapes in these EU countries would imply the creation of 119,000 permanent jobs in the MPCs, with total economic profits to producers and exporters being in the order of about US$498 million and added value to the national economies of the five MPCs equalling US$756 million. The social and economic impacts were predicted by translating the unmet demand from metric tons into hectares and then using input-output coefficients to calculate the labour requirement, the permanent jobs, the social added value and the water requirement. The authors noted that structural change and adjustments, foreign investment and changes in the legal framework need to take place in the MPCs to make the conditions favourable for further exploitation of their comparative advantages. One future constraining factor could be water supply, which is becoming the most limited factor in terms of volume and quality.

A project carried out within the Femise network examined the development of the Euro-Mediterranean Partnership focusing on the existing impacts on trade in industrial commodities and services and the impacts on the labour market stemming from

---

[12] The selected crops are green house tomatoes, green house green beans, strawberries (both green house and open field), table grapes, sweet melons, green house sweet peppers, apples, dates, olives, anises and roses.

flows of investments aiming at creating structural changes (HABDOUSSA and REIFERS, 2003). After analysing the current macroeconomic environment of the MPCs the authors identified the areas – axes for development over a fifteen-year horizon. They suggest that the focus should lie on creating job through undertaking necessary structural adjustments in the labour market and education system and assuring macroeconomic stability. Given the current labour market structure and population growth in the coming fifteen years, 34 million new jobs are required to avoid any decline from the present situation. The authors suggest that macroeconomic stability requires reform of the taxation system, a reduction of public expenditure and controlling inflation. The MPCs should prioritise the objectives of expanding economic growth, modernising the infrastructure, improving trade competitiveness and addressing the issue of poverty. Looking at the poverty level, although only 2 % of the population live on less than a dollar per day, about 30 % of the population (72 million people) live on less than 2 dollars per day, a rate identical to this of 1987. The key to accomplishing these objectives is attracting foreign capital. Despite the Euro-Med Agreements, European investors seem to prefer to invest within EU candidate countries or South America rather than in the MPCs. American investors appear to be more attracted to the region.

Gravity models

GARCIA ÁLVAREZ-COQUE and MARTÍ SELVA (2006) have assessed the effects of the Euro-Med Agreements on the trade of fresh fruits and vegetables between the EU and the MPCs by using a gravity model. The gravity equation shows the extent to which bilateral trade flows are determined by variables that indicate the total potential demand in the importing country, the total potential supply of the exporting country and by binary variables that reflect the effects of the Association Agreements between the EU with the MPCs in increasing the value of trade between the two. The model was applied to the EU-15 and eight south Mediterranean countries, Morocco, Algeria, Tunisia, Egypt, Jordan, Syria, Lebanon and Israel, over the period 1995-2004. The results indicate that the decisive factors for the trade of fresh fruit and vegetables between the EU and the MPCs are the distance between the exporting countries and the EU, the demand patterns in the EU and the cultural and historical links between particular exporting countries and the EU. In spite of logistical and technological improvements, the proximity to EU markets has a considerable influence on trade and provides as a significant premium for retailing companies. Overall the Association Agreements have boosted the exports of fruit and vegetables from the MPCs to the EU, but when looking at individual countries, the study shows that the Agreements have only marginally helped to integrate the Maghreb and Mashrek economies and their effect has only really been significantly beneficial for Israel. The authors argue that this could be explained by the growing demand for quality products that is better fulfilled by more developed economies, such as Israel, rather than Maghreb and Mashrek countries.

Ex ante studies

The existing *ex ante* empirical studies focus on analysing the impacts of future trade liberalisation between the EU and the MPCs. Most of them have been carried out using CGE models and only a few with PE models. One study employed an Input-Output (I-O) model and another study used a non-linear optimisation model. The following section gives more insights on each of these empirical studies.

Equilibrium Analyses

Computable General Equilibrium (CGE) modelling has been widely used to analyse the impact of the Euro-Med Agreements, since it offers the advantage of capturing economy-wide effects. Nevertheless, because of the extended database required to support such a model, most of the studies focus only on one country, usually Turkey, Egypt, Tunisia or Morocco, using national CGE models. A number of studies that employ multi-regional, multi-commodity models have used the GTAP model without modifying its structure or closure, or alternatively use the database from various GTAP versions.

A few studies have been carried out with dynamic CGEs, and fewer still have used PE models to analyse the impacts of trade liberalisation on the Mediterranean agricultural sector.

An overview of the empirical studies and their scope follows in Table 4.1.

Reviews of the empirical studies which model the Euro-Med Association Agreements and regional and preferential agreements (including the Euro-Med) are provided by KUIPER (2004), KURZWEIL et al. (2003) and NIELSEN (2003). In addition, an extensive literature review on the sustainability of the impacts of the Euro-Med Agreements has been prepared by the consortium of the SIA project (Sustainability Impact Assessment) (IARC, 2005 and 2004).

Table 4.1: Overview of ex-ante empirical studies on modelling agricultural trade policy liberalisation on the Mediterranean Basin with equilibrium models

Type of model	Study	Scope of the study
Computable General Equilibrium Models		
static	AUGIER and GASIOREK (2003a)	Euro-Med Agreements
	BROWN et al. (1997)	EU-Tunisia free trade area
	CHATTI (2003)	EU-Tunisia free trade area
	HARRISON et al. (1997)	Turkey customs union
	HOEKMAN et al. (2001)	Egypt's trade liberalisation
	KONAN and MASKUS (1997 and 2000)	Egypt's trade liberalisation + fiscal policies
	LUCKE (2001)	Jordan's and Syria's fiscal impact of trade liberalisation
	MINOT et al. (2007)	Impacts of trade liberalisation on Tunisia and Syria
	RAVALLION and LOKSHIN (2004)	Moroccan trade liberalisation
	RUTHERFORD et al. (1997)	EU-Morocco free trade area
- GTAP	ALESSANDRI (2000)	Euro-Med Agreements
	DENNIS (2006)	Euro-Med Agreements/GAFTA
	DIAO and YELDAN (2001)	Euro-Med Agreements
	ELBEHRI and HERTEL (2004)	EU-Morocco free trade area
	HOSOE (2001)	Jordan trade liberalisation
	KUIPER (2006)	Euro-Med Agreements on Morocco and Tunisia
	MERCENIER et al. (1997)	EU-Turkey customs union
	SONMEZ et al. (2006)	Turkey accession to the EU
dynamic	BOUËT (2006b)	Trade strategies of the MPCs towards the EU
	CHEMINGUI and DESSUS (2001)	EU-Tunisia trade liberalisation
	FERABOLI et al. (2003)	EU-Jordan trade liberalisation
	LÖFGREN et al. (2001)	Moroccan trade liberalisation
Partial Equilibrium Models		
static	AUGIER and GASIOREK (2003b)	Euro-Med Agreements on Morocco
	BRITZ et al. (2006)	Euro-Med trade liberalisation
	GARCIA ÁLVAREZ-COQUE et al. (2007)	Euro-Med Agreements on Moroccan tomato market
	GRETHE (2003)	EU-Turkey customs union
spatial	M'BAREK (2002)	Euro-Med Agreements on Morocco and Tunisia

Source: Own compilation

Static CGE models

HARRISON et al. (1997) developed a CGE model to quantify the impacts of the customs union between Turkey and the EU. Seven scenarios were developed in total, where various policy adjustments that Turkey has to undertake are modelled. These include: a tariff reduction on Turkish imports of manufactured products so as to comply with the EU Common External Tariff; the impacts of improved market access only to EU markets; improved access because of harmonisation of product quality standards; improved market access to the markets of EU's preferential trade partners; a reduction of export subsidies; a reduction of trading costs and finally; the overall customs union. The gains for Turkey vary between 1 % and 1.5 % of its GDP and compensate the losses from tariff revenues. However, the authors do not report more on the model or the dataset used for the simulations. They indicate that a broader model documentation is included in HARRISON et al. (1996) (a publication that could not be accessed).

MERCENIER and YELDAN (1997) examine the customs union between Turkey and the EU and analyse the implications of trade liberalisation on the agricultural sector by using an intertemporal CGE model. The model recognises increasing returns to scale, the existence of firm level product differentiation and an oligopolistic market structure. The demand side is covered through a single representative household. The model entails nine sectors, four of which are perfectly competitive (agriculture and primary products; food, beverages and tobacco; other manufacturing industries and; transport and services) while the other sectors are considered to be non-competitive (pharmaceutical products; other chemistry; motor vehicles; office and other machinery and; transport materials). It divides the world into seven regions, i.e. Turkey, Great Britain, Germany, France, Italy, the rest of the EU-12 and the rest of the world. These are linked by an Armington system. In two scenarios the authors simulate Turkey's commitment to entering into a customs union with the EU. In the first they assume the elimination of tariffs on European imports and harmonising the tariff rates with the rest of the world with the existing EU ones. In the second they assume the possible accession of Turkey into the EU with a single pricing system. The benchmark used for the simulations is a '92 post-Europe computed equilibrium. The authors only find welfare gains, of just below 1 % of GDP for the Turkish economy, in the event of the second scenario. They conclude that a simple harmonisation of Turkey's tariff system would result in a loss of real consumption of almost 1 % and would only be beneficial if accompanied by full trade integration.

RUTHERFORD et al. (1997) applied a general equilibrium model to examine the effects of trade liberalisation between Morocco and the EU. The model is based on a previous study of RUTHERFORD et al. (1993) and belongs to a category known as small open economy models. It has one representative consumer, assumes no terms-of-trade effects, no capital accumulation and constant returns to scale with competitive pricing. The authors use a two-tier CET function to firstly determine the split between domestic sales and composite export sales and secondly to distinguish between exports sales to the EU and non-EU countries. The Moroccan economy is organised into 19 sectors, seven of which are agricultural (cereals, sugar, citrus fruits, vegetables, meat and dairy, fishing, forestry and other agriculture). The model utilises production and value-added data from a SAM of 1980 and tariff data of 1991. Six scenarios are simulated for various trade liberalisation options that range from improved market access for Moroccan fruits and vegetables in the EU to a unilateral full trade liberalisation of Morocco with the rest of the world. It is shown that the EU-Morocco free trade area will increase welfare in Morocco by about 1.5 % of GDP, with the effects being higher with full unilateral liberalisation (2.6 % of GDP). They find the trade creation effects are large than the trade diversion effects. Key parameters used to identify whether a regional trade agreement results in net positive results or not, include: import shares, tariff levels and the substitution elasticities in consumption. The results indicate that a higher Armington elasticity leads to higher welfare benefits. The study finds a correlation between welfare effects and the required factor adjustments, but shows that broader trade liberalisation creates more welfare gains than a free trade area but in this case the adjustment costs slightly higher.

BROWN et al. (1997) study the impacts of the free trade area between Tunisia and the EU, using the "Michigan Brown-Deardorff-Stern CGE trade model". The model assumes full employment, fixed relative wages and a fixed labour supply. The agricultural sector is assumed to be perfectly competitive, and all other sectors are assumed to be monopolistically competitive. Agricultural commodities are differentiated, according to their origin and firm differentiation is applied to all other commodities. Of particular interest is the regional aggregation that is followed, with the EU-12 divided into three regions, one containing Greece, Portugal and Spain, France and Italy a second and the rest of EU-12 a third region. Other regions included in the model are Tunisia, the rest of Europe, Asia and the Pacific, the North Atlantic Free Trade Area (NAFTA) and the rest of world. The simulated scenarios involve the bilateral removal of tariffs and non-tariff barriers on goods between Tunisia and the EU-12, combined with sectorally mobile capital and a flow of capital from the EU into Tunisia. The results show that the welfare benefits for Tunisia range from slightly negative (-0.2 % of GDP) to somewhat positive (+3.3 %) depending on the sectoral mobility of capital. The welfare gains remain almost unaffected by capital flows.

KONAN and MASKUS (1997) examine the impact of trade liberalisation on the Egyptian economy, which is modelled as a small open economy that trades differentiated goods and services with multiple regions. The imported and domestic goods are imperfect substitutes as are domestic goods and exported ones. The demand side is covered by using a single representative household, while changes in aggregate consumption are used as direct measure of the equivalent variation of a policy change that is also used to measure welfare changes. Benchmark data is taken from a SAM for Egypt from 1990, which the authors updated to 1994. Three of the sectors covered in the SAM represent the agricultural sector, namely vegetable food, vegetable non-food and animal products. The model divides the world into five regions, namely Egypt, the EU and Turkey, the USA, the rest of Middle East and North Africa, and the rest of the world. The authors use five scenarios to simulate various options for liberalising trade, and throughout these they assume that Egypt adjusts its domestic tax system to compensate for a reduction of tariff revenue. The results show that Egypt would enjoy significant welfare gains of about 2.0 % of the GDP from reducing red-tape costs on imports and exports, which are taken into account through an assumed tariff equivalent. With this reform, the association agreement with the EU would slightly reduce welfare. The welfare gains are higher when trade liberalisation becomes broader and the highest welfare gains are observed in the scenario where full unilateral liberalisation of trade is simulated (about 3.0 % of GDP). In a follow-up study the same authors focus on the interactions between trade liberalisation and changes in domestic fiscal policies (KONAN and MASKUS, 2000). They conclude that Egypt could enjoy welfare gains of to 1 % from reforming its tax system. Trade liberalisation also produces welfare gains, but lower due to trade diversion effects and lower tariff revenues. A combined effect would result in welfare gains depending on the magnitude of each reform.

ALESSANDRI (2000) uses the GTAP model, version 4, to assess the impacts of the EU trade policy under the framework of the Barcelona Agreement. The simulations rest on a 10-regions-10 industries aggregation of the GTAP database. The agricultural sector is represented through crops and other agricultural products. The study examines the customs union between the EU and Turkey, the Euro-Med Agreements between EU and Morocco and the Euro-Med Agreements between the EU and the rest of North Africa which are modelled as reciprocal elimination of trade barriers (import tariffs) on manufactured goods. Only two scenarios are relevant for the agricultural sector, in one import tariffs are eliminated and in the other import tariffs are eliminated and output and export subsidies are dismantled. The findings suggest an increase of the welfare varying from US$3.3 to 2.6 billion for the EU, but an ambiguous impact on the MPCs. For example in Morocco the welfare increase varies between US$0.38 and 0.47 billion (the higher figure is for the elimination of both import tariffs and output-export subsidies) and in Turkey of about US$0.82 billion (for all scenarios). For the rest of the North African countries the welfare deteriorates slightly (welfare decreases of US$0.23 billion).

DIAO et. al. (2001)[13] extended and adjusted the same model used by MERCENIER (1997) to examine the static and inter-temporal effects of bilateral trade liberalisation between the EU, Turkey and non-EU Mediterranean countries. The model is based

---

[13] The paper has been also published as BAYAR et al. (2001 and 2000).

Empirical studies

on inter-temporal general equilibrium theory with Ramsey-type dynamics. The world is divided into nine regions with EU, Turkey, Morocco, the rest of the Middle East and the rest of North Africa as separate regions. It includes nine sectors and the products are differentiated according to their geographical origin (Armington assumption). Two simulation scenarios are developed, with the customs union between Turkey and the EU used as the baseline scenario. The first scenario simulates trade liberalisation of the manufactured goods between the MPCs and the EU and in the second this liberalisation is extended among the MPCs themselves. The data come from simple aggregation of the GTAP version 3 database. Manufacturing trade liberalisation entails static welfare losses for the MPCs and welfare gains for the EU in both scenarios, with the losses being smaller in the second scenario. Nevertheless, when dynamic aspects such as investment and growth effects are taken into account, then the model predicts welfare gains in all regions. The authors conclude that the static effects are the short term effects, with the welfare loses explained by a deflation of domestic prices in the MPCs and vice versa.

The impacts of trade liberalisation scenarios between Egypt, the EU and other Arab countries are examined by HOEKMAN et al. (2001) using a standard, single country, competitive, computable general equilibrium model. The model entails 38 sectors, three of which refer to agriculture. Three preferential trade liberalisation scenarios are modelled. The first refers to removal of import tariffs on EU goods and improved access to the EU market for Egyptian products (through higher export prices of 1 % for the Egyptian products), the second looks at the additional removal of tariffs on commodities originating from the USA and the third, the additional deeper integration with the USA through elimination of all trade barriers. The results indicate welfare gains of 0.99 %, 1.26 % and 2.31 % of GDP for the three scenarios respectively. The highest effects for Egypt are achieved through the formation of a free trade area with the USA, but this is to the cost of the rest of the world due to trade diversion effects.

A comparative static national computable general equilibrium model has been developed by HOSOE (2001) to analyse the impact of the Uruguay Round and the Euro-Mediterranean Association Agreements on Jordan. The model is calibrated to the GTAP database version 3 and input-output tables of Jordan. The model includes nine sectors and three regions (Jordan, EU and the rest of the world). The agricultural sector is represented by a single sector. The Uruguay Round scenario consists of tariff cuts and a phasing out of the Multi fibre Arrangement, whereas the scenario of a free trade area with the EU assumes a reduction of tariff rates between Jordan and the EU by 80 %. Social welfare is measured by the Hicksian equivalent variation as the relative size of the base run GDP. In both scenarios Jordan's welfare is improved, by 0.28 % and 0.16 % respectively. This change is attributed to trade creation effects in exports and imports, changes in sectoral output and favourable trade diversion effects favourably on imports from the EU.

LUCKE (2001) developed a CGE model to analyse the fiscal impacts of trade liberalisation for Jordan. The entire economy of Jordan is decomposed in 13 sectors including agriculture (aggregated as one sector). The world is broken down into four regions, namely Jordan, the Middle East and North Africa, the EU and the rest of the world. The model uses constant elasticity of substitution (CES) and constant elasticity of transformation (CET) functions to determine supply and demand, thus distinguishing domestically produced commodities from imported ones. The model is calibrated using a 1998 SAM for Jordan, with data from missing elements taken from a 1987 Input-Output table. The possible effects of the Euro-Med Agreements are tested in seven scenarios, which simulate various degrees of trade liberalisation. The results indicate that a reduction of import tariffs causes a chain of effects starting from reduced domestic prices for imported goods, consumption of domestically produced commodities being substituted by imported ones, an increase of imports, a decline of domestic prices for domestic products and a switch from the domestic supply of goods to exports. Real GDP slightly decreases in the full liberalisation scenarios. The nominal and real effects of only liberalising trade in agricultural products are rather negligible. The losses of government revenue, due to reduced tariff revenue, are about 14 % under full liberalisation. The government deficit though remains almost stable because government consumption and investment decline but in the same proportion as the fall in government revenue. The rise in the current account deficit is much higher (for example 717.9 % under the full liberalisation scenario which amounts to 8 % of nominal GDP), reflecting the rise in the trade deficit. The losses in tariff revenue could be compensated for by cutting real government consumption by about 5 % and real public investment by about 3 %. It should be noted that because all the prices have been normalised in the base run scenario to one hundred and also because the consumer index value is used as numéraire with a value of one hundred throughout the simulations, the counterfactual values of prices can only be seen as index values relative to the consumer price index.

A CGE for Syria was developed in the same study (LUCKE, 2001). Benchmark data for this model used a SAM for Syria from 1999, with the economic activities being aggregated into nine categories (one of which covers agriculture). The world was aggregated into nine regions (Syria, Arabic States, EU-15, ex-socialist countries, USA, Turkey, Japan, ABC – Argentina, Brazil and Chile - and the rest of the world). A composite good was produced for each product category, using the inputs, domestic supply and imports in a CES function. Trade creation is described as a function of the relative price between domestic and imported commodities. Trade diversion is modelled assuming that, for a given import volume, Syria minimises the costs of imports with trading partners under CES technology, and CET transformation functions are used to model aggregated export. The picture in Syria is very similar to that of Jordan. The results of seven liberalisation scenarios show a gradual increase in producers' prices and an increase in the trade deficit due to an increase in imports. The increase in the total government deficit is almost equal to the loss of tariff revenue. Due to trade diversion effects the EU is the trade partner benefiting the most. The author argues that Syria could offset the tariff revenue losses through indirect taxation on consumption goods thereby minimising any problems arising from the socially unacceptable distribution of the tax burden.

AUGIER and GASIOREK (2003a) carried out a non-GTAP study. They built an 11 country – 10 sector static CGE model, allowing for imperfect competition in product markets and increasing returns to scale in production, to examine the price and welfare implications of liberalisation between the EU and southern Mediterranean countries. The scenarios involve full liberalisation (i.e. elimination of all tariffs) as well as improved market access and trade-induced changes in productivity. The welfare effects are measured by a compensating variation as a proportion of base GDP. Under the full liberalisation scenario all countries gain due to trade creation except for Jordan, Syria, and Turkey, whereas the highest welfare gains are for Tunisia (8.9 %), Morocco (5.36 %) and Egypt (1.39 %). The tariff-reduction and the improved market access and productivity scenario results in welfare gains for all the southern Mediterranean countries while the effects are rather minor for the EU. The results are similar when reducing the tariffs levied by the southern Mediterranean countries to the EU level (i.e. to Most Favoured Nation level). The results do not change when the tariff reduction gradually takes place, a scenario closer to the framework of the Euro-Med Agreements.

CHATTI (2003) examined the impacts of the Free Trade Agreement between Tunisia and the EU assuming both a competitive market and oligopolistic market structures. Three scenarios were simulated with three different versions of a static single country CGE model, based on the model provided in RUTHERFORD et al. (1997). The model used Tunisian data from 1995 and includes 22 sectors, with agriculture aggregated into one sector. The basic version assumed constant returns to scale, perfect competition and exogenous terms of trade, while the commodities are differentiated for both imports and exports. The second and third versions of the model introduce increasing returns to scale through positive fixed costs and constant marginal costs, respectively. The firms behave oligopolistically, meaning that the unit costs are greater than the marginal ones and therefore the marginal cost pricing rule does not apply. Nevertheless it is assumed that Tunisian firms only enjoy market power in the local market and face an infinitely elastic export demand. These two versions respectively apply either prohibitive or free entry and exit to the oligopolistic industries. The welfare effects are measured through the equivalent variation as a percentage of GDP in 1995. The results under perfect competition and constant returns to scale (first model version) show that Tunisian welfare increases by 6.5 % of GDP, due to preferential trade liberalisation of manufactured goods from the EU, with trade creation being higher than trade diversion. The welfare gains are higher under the second and third versions of the model, at about 7.1 % and 7.3 % respectively. The author notes that in all cases the FTA seems to benefit manufacturing sectors which expand, but at the expense of agriculture and private services.

ELBEHRI and HERTEL (2004) employed the GTAP model, version 6.1 to examine the Morocco-EU Free Trade Area versus multilateral liberalisation. The world is aggregated into three regions, Morocco, the EU and the rest of the world. The Moroccan economy is divided into 28 sectors, 15 of which describe manufacturing activities and nine agricultural activities. The underlying data are from the GTAP version 5.3 database and incorporate a SAM of Morocco of 1990 into the GTAP database. Unilateral liberalisation results in a deterioration of the terms of trade for Morocco and thus in welfare losses (a decrease of US$392million, $16million and $1935.9 million respectively when no entry, entry and no entry and unemployment are assumed). The effects on the output per firm in industries dominated by scale economies and the effects on the aggregate demand for labour are adverse, with the imports being diverted to non-EU suppliers. However, multilateral liberalisation under the WTO Doha Round, realised through a 30 % across the board cut in all tariffs, results in welfare gains for Morocco (US$414.6 million, $528 million and $666.4 million respectively under no entry; entry; and no entry and unemployment), which is attributed to lower losses from the terms of trade, positive scale effects, a positive impact on the labour demand and non-preferential imports into Morocco.

The study of RAVALLION and LOKSHIN (2004) deals with the analysis of the impacts of trade policy reforms in Morocco. The authors use a static CGE calibrated to 1997 to simulate trade liberalisation and to estimate the price changes and welfare effects. The model was initially designed by World Bank and the Moroccan Ministry of Agriculture to assess the aggregate impacts of removing protection from cereals in Morocco[14]. The model entails 16 different crops, three livestock activities, 13 major agro-industrial activities, six agro-ecological regions and within each region a distinction is made between rain-fed and four types of irrigated agriculture. Two types of labour have been included with fixed real wage rates. However, the authors do not specify the crops, the other activities or these two labour types. In four simulations, tariff cuts of 10 %, 30 %, 50 % and 100 % are undertaken, and in all the scenarios Moroccan open-market operations are removed (which subsidise sales of cereals so as to keep consumer prices at a low level). The results indicate that full trade liberalisation would lead a reduction by 24 % of the grain producers' price and to a reduction by 27 % of the price paid by consumers for grain products, while the effects on other crops are indirect and are less than +/- 3 %. In a second step the authors combine the price changes with household survey data from 1998 covering 5,117 households in 14 of

---

[14] Details on this modelling exercise are documented in „DOUKKALI, R. (2003). Etude d'Effets de la Libéralisation des Céréales: Resultats des Simulations à L'Aide d'un Modèle Equilibre Général Calculable. Joint Report of the Ministry of Agriculture and the World Bank" This source could not be found and information on this modelling exercise is based solely on RAVALLION and LOKSHIN (2004).

the 16 regions in Morocco. These data are used for calculations, based on standard first-order welfare analysis, to estimate the gains and losses of the induced price changes at the household level. By using regression analysis they aggregate the results and calculate changes in the mean income and the incidence of poverty. They estimate partial trade liberalisation reforms to slightly increase the national poverty rate, with the effect being larger under full trade liberalisation, where the poverty rate rises from 20 % to 22 %. All four reforms decrease urban poverty and increase rural poverty. Although the net consumers of cereals benefit from lower prices, the effects on rural poverty arise because, among the poor in rural areas, the losses of the net producers do not outweigh the gains of the net consumers.

KUIPER (2006) employed version 6 of the GTAP model to examine the possible outcome of the implementation of the Euro-Med Agreements on the economies of Morocco and Tunisia. The regions of the GTAP database are aggregated into nine blocs, where Morocco, Tunisia, northern and Mediterranean EU member states being modelled separately. The primary agricultural commodities are aggregated into cereals, oilseeds and vegetable oils, vegetables-fruits and nuts, spices and other crops, plant based fibres, sugar cane, sugar beet, animal and wool products, milk and dairy products. The model closes assuming unemployment for unskilled labour instead of the standard assumption of perfect labour markets. Four simulation scenarios are examined, which include extreme situations providing the upper and lower bounds of the implementation of the Euro-Med Agreements. The first two scenarios simulate the current implementation of the Agreements through the elimination of the tariffs for manufactured commodities, while the next two scenarios simulate full liberalisation, i.e. eliminating the tariffs on agricultural commodities as well. In each of these two simulations another one is undertaken, where the tariffs are replaced by a consumption tax to make up for the losses to the budget of the states due to reduced tariff revenues. The baseline scenario includes the implementation of the remaining Uruguay round commitments for developing countries that end in 2005 together along with phasing out the restrictions on trade in textiles from 2005 onwards (Agreement on Textiles and Clothing), China's WTO accession, the EU enlargement in 2004 and the latest CAP reform (Luxembourg Agreement). The results indicate that the northern EU member states would benefit from improved market access to North Africa countries for cereals, animal and dairy products, while the Mediterranean EU member states would face an increase in imports of vegetable oils. The elimination of tariffs on manufactured commodities reduces distortions in Morocco and Tunisia and diverts trade from the rest of the world to the EU. The income gains for Morocco are the highest (4.84 % of GDP in the base run) and are relatively insignificant for the northern and southern EU regions (0.01 and 0.02 % of GDP in the base run respectively). The replacement of tariffs with a consumption tax leads almost to no further changes for the EU but to a decline of welfare in both Morocco (-2.63 % of GDP in the base run) and in Tunisia (-1.64 %). Scenario 3, which includes liberalisation of the agricultural sector without tariff replacement, leads to additional gains of 1.2 % of the GDP of the base run in Morocco and to 4.8 % of the GDP in Tunisia. Tariff replacements lead to minor income losses of 0.2 % of the base run GDP in Morocco and to income gains of 1.1 % in Tunisia. One notable effect of the full liberalisation scenarios is a spectacular increase of the vegetable oil production in the MPCs. Since in the base run most vegetable oil exports (82.1 %) are destined to EU Mediterranean countries, this implies that these regions will be the most affected by the liberalisation of agricultural trade. Vegetable oil production in Tunisia shows an eightfold increase. The production of cereals declines, with and without tariff replacement, and EU countries benefit from this. No conclusions for fruits and vegetables can be drawn due to the high level of aggregation of these commodities in the model.

Using the version 6.0 of the GTAP database SONMEZ et al. (2006) employ the general equilibrium model Globe to assess the impacts of Turkey's accession in the EU. They use a 15-region, 15-sector and 4-factor world trade CGE, with agriculture aggregated into one sector. The regions are the EU-15, broken down to Germany, France, UK, Italy and the rest of the EU, Switzerland, the rest of Europe, Russia, the rest of the former Soviet Union, Turkey, the rest of the Middle east, Japan, USA and the rest of the world. The model is SAM-based, where production follows a two stage nest structure, the final demand of the government and of investment is modelled under the assumption that the demanded relative quantities of each commodity by these institutions are fixed. The utility function of private households is assumed to be of the Stone-Geary type and the commodities are distinguished between domestically produced and imported ones. The basic model closure is a full employment balanced macroeconomic closure with unemployed unskilled labour in some regions, where assumptions such as flexible exchange rates, fixed shares of investement expenditures in final demand, fixed tax rate adjusters, mobile and fully employed factors and regional specific consumer price indeces to serve as regional numéraires are made. In this application two variants on the closure were run with the first assuming only full employment so as to assess the effects of assuming unemployed skilled labour in Turkey and the second a combination of the first with a balanced macro closure. The simulated scenarios involve a 100 % removal of import tariffs and export duties on all traded commodities, taking place one at a time and simultaneously and a 100 % liberalisation together with Turkey adopting the EU's Common External Tariff system. The base year for the simulation is 2001, when the trade in industrial and processed agricultural commodities was already fully liberalised, due to the customs union between Turkey and the EU. Therefore, the removal of custom duties involves only non-processed agricultural commodities within the developed scenarios. Liberalisation leads to an increase in GDP from value-added of about 11 % when only import duties are removed, by only 2 % when only the export subsidies are eliminated and by about 20 % when done together and with full liberalisation (imposing the Common External Tariff). The import demand is expected to increase by 32 % with the elimination of import duties and by 52 % with the elimination of all custom duties. Figures for the export supply under these two assumptions increase by 12 % and 28 % respectively. The tariff revenue decreases by between 8 % and 41 % depending on the scenario. There are similar decreases in export tax revenue. The domestic production of agricultural commodities increases by about 7% as a result of full liberalisation. The exports of agricultural products to countries other than the EU decrease under the balanced macro closure whereas, with the exception of Russia and the rest of the Middle East, they increase under the unemployment labour closure. Overall the effects are more profound under the unemployment labour closure.

DENNIS (2006) examined the impact of regional trade agreements and further trade facilitation on the Middle East and North Africa (MENA) regions. The analysis is carried out using the GTAP model, version 6 incorporating trade facilitation by splitting trade transaction costs into indirect trade transaction costs ("iceberg" costs) and a tax component capturing the direct trade transaction costs. The study uses a 13-region, 16-commodity aggregation of the GTAP database, with the Mediterranean countries aggregated into two groups, the rest of North Africa and the rest of the Middle East, apart from Morocco and Tunisia which are modelled separately. The agricultural sector is broken down to fruits and vegetables and fish and live animals. Two simulation scenarios are formulated, one where all import tariffs between Middle Eastern and North African countries are abolished (a scenario that reflects the GAFTA Agreement) and one where, in addition to this the import tariffs for non-agricultural commodities between the EU and the MENA countries are reduced by 100 % and those for agricultural commodities by 50 % (a scenario reflecting the Euro-Med Agreements). For both simulations a counterfactual simulation, with and without trade facilitation improvements, was undertaken. The results indicate welfare gains for the MENA region amounting to US$913 million under the first scenario and US$1.84 million under the second, clearly showing the importance of integration with the EU. Nevertheless there are regional differences in the distribution of the welfare gains among the MENA countries. For example in the scenario of liberalisation between the MENA countries Tunisia gains the most and enjoys welfare increase of +0.53 % followed by the rest of Middle East, whereas in the scenario with the EU-MENA liberalisation Morocco enjoys the highest welfare gains of about 1.88 % followed by Tunisia (welfare gains of about 1.72 %). The differences in the results between the MENA regions are analogous and proportional to their trade expansion. The changes in GDP follow the same direction as the welfare changes i.e. the real GDP rises in all MENA regions under both simulations, varying from 0.02 % to 0.21 % for the first simulation and 0.12 to 2.22 % for the second. Again Tunisia enjoys the highest benefits under the first simulation while Morocco and Tunisia gain the most under the second scenario. Because full employment closure of the model is adopted the increase of GDP points to an increase in economic activity and positive effects the factor markets, which in the case of real wages can be translated into increased employment.

The results are far more profound when trade facilitation improvements are modelled. For example by running the first scenario with the trade facilitation option, the welfare gains for the MENA countries are at least three times higher than without trade facilitation. A decomposition of the welfare gains reveals that they are mainly gains in allocative efficiency, while further decomposition reveals that trade facilitation improvements from lowering indirect costs contribute more to the welfare gains than lowering the direct trade transaction costs.

MINOT et al. (2007) have done two case studies examining the impact of trade liberalisation on agriculture and poverty. The first one, of Tunisia, uses a static CGE which is based on a Tunisian SAM from 1996. It incorporates data from a household expenditure survey among 400 households in 1995. The authors assume perfect mobility of physical capital among different sectors and imperfect substitution among domestically and non-domestically produced commodities. Supply is modelled using a nested CES function and substitutions among labour categories are implemented through nested CES functions for agricultural and non-agricultural activities. The commodities are aggregated into 15 categories, including grain crops, leguminous crops, fruits, vegetables, meat, fishing products, dairy products, sugar, olive oil and tobacco products. The first scenario simulated the elimination of tariffs on imported industrial products from the EU. In the second the tariff elimination is extended to agricultural products and, in the third, the elimination of tariffs on industrial and agricultural products is also extended to the rest of the world. The assumptions of the third scenario are carried over into the fourth scenario and combined with an increase in world market prices for all agricultural products by 15 %. Openness in trade would result an expansion of imports and exports but almost exclusively in non-agricultural sectors. There are hardly any effects on the agricultural commodities from the first scenario, while under the second scenario imports of meat, fruits and dairy products would expand by about 164, 63 and 60 % respectively. The increase of exports of grains would be the highest (about 50 %). The effects of the third scenario are similar to those of the second scenario, while under the fourth scenario a further increase of exports occurs, especially of dairy products and of sugar (about 130 % respectively). The changes in poverty are only marginal.

In their second study, MINOT et al. (2007) examine the impact of liberalisation of the wheat market on small farmers in Syria. They used a static CGE developed by LÖFGREN et al. (2001), which is built using a Syrian SAM from 1999 and includes 10 representative households for each decile of the Syrian population. The model simulates to a reduction in the domestic subsidies for wheat production and consumption of 20 and 50 % and 100% respectively. The macroeconomic effects are rather moderate and government savings increase by up to 3 % throughout the scenarios. Complete liberalisation leads to reduction of producers' prices by about 17 % and about a 2% decrease in production. The effects of subsidy removal on the welfare of the Syrian households are regressive, with high-income households gaining and lower-income households losing, although they are of a small magnitude (less than 1 % of the base income for all households).

Dynamic CGE studies

CHEMINGUI and DESSUS (2001) created a dynamic CGE to model sequential tariff cuts due to liberalisation in trade between Tunisia and the EU. The model takes two representative households, one rural and one urban and additionally a tourist household. It includes 57 sectors, 26 of which are related to agriculture or food industries, and distinguishes two trading partners for Tunisia, i.e. the EU and the rest of the world. It is calibrated using a SAM from 1992 for Tunisia. The model assumes imperfect competition among goods coming from different geographic areas and is resolved recursively every three years from 1992 to 2010. The baseline scenario involves growth of the Tunisian economy and the implementation of the GATT, involving tariff cuts and the abolition of non-tariff barriers as well as a progressive reduction of tariffs on industrial products, as foreseen by the Euro-Med Agreements. The scenarios refer to unilateral liberalisation i.e. tariff reductions or the abolition of non-tariff Tunisian measures towards imports from the EU or less governmental support for agricultural products. The results indicate a worsening of Tunisia's agriculture (although the trade volume increases) since domestic production has to compete with commodities imported from the EU. The welfare effects are of a low magnitude, they are positive for the rural households only in the events of increased access to the EU for Tunisian agricultural exports (+2.7 %) and multilateral liberalisation (+0.2 %). The aggregate welfare effects throughout the simulations are slightly positive and higher when liberalisation is multilateral (+2.5 %).

LÖFGREN et al. (2001) have developed a dynamic recursive computable general equilibrium model of Morocco to analyse alternative policy scenarios[15]. The model uses CES production functions, and substitutability is assumed based different geographic origin. The model has a savings-driven determination of investment and this is solved as a mixed-complementarity problem. It distinguishes between rural and urban activities and households, uses data from a Moroccan SAM from 1994 and it has a detailed representation of the agricultural sector. Of the 45 activities, 38 are rural and most of them relate to agricultural or livestock products. The static model is solved for 1994 and 1998 and the solution from the 1998 run is used to update the base year run. Following this step the model is solved for every two years up to 2012. Between the solutions selected parameters are updated in the dynamic module. The baseline simulation involves a gradual implementation of the EU partnership. While the first scenario assumes unilateral trade liberalisation, the second combines unilateral liberalisation with changes in domestic policy regarding rural education and the imple-

---

[15] A detailed description of the model is given in LÖFGREN et al. (2002).

mentation of a non-distorting programme for cash compensation to owners of resources used in rainfed agriculture. The results indicate that tariff unification has rather limited impacts on aggregate factor incomes and household welfare. The later is the highest under the second scenario and is not greater than 2.5 % of 1994 GDP. The removal of trade barriers and the reduction of tariffs lead to major expansion of non-agricultural exports, significant growth of non-agricultural sectors and a slow down in the agricultural ones. Trade liberalisation seems to reduce income growth for agricultural resources and disfavours poor rural households, but when combined with complementary domestic policies it can improve the welfare of all household types.

A simple dynamic CGE has been developed by FERABOLI et al. (2003)[16] to examine the impact of a free trade area between Jordan and the EU. The authors developed a simple neo-classical open economy model with two production sectors, goods and services, assuming perfect competition and full employment in both sectors. The model allows for imperfect substitution between domestic and foreign goods. The demand side is covered though a single representative household, the model closure is based on fixed current-account balance and the saving-investment is solved through inter-temporal maximisation. The dataset is based on a SAM of 1998 developed by LUCKE (2001), but the model is further simplified by aggregating all goods' sectors. The dynamic nature of the model allows liberalisation to be modelled by gradually reducing the tariff over a period of 13 years, as foreseen by the Association Agreement between Jordan and the EU. Two sets of simulations were carried out, involving the implementation of the free trade area between Jordan and the EU and a non-discriminatory trade openness by Jordan. The results show that non-discriminatory tariff reduction has more effect in increasing GDP and labour demand in the long run than the Association Agreement with the EU. The welfare effects are positive throughout the scenarios but the magnitude of them depends on the fiscal measures adopted by the government to counteract for losses of the tariff revenue due to the progressive reduction of tariffs. When capital tax is endogenous and labour income fixed then the free trade area with the EU appears to be more welfare-enhancing than under broader liberalisation. In the scenarios where the labour income tax rate is endogenous and the capital tax is fixed, trade openness yields larger welfare gains than the free trade area. Generally the authors suggest that investment incentives are higher under the preferential agreement with the EU.

The MIRAGE model has been employed by BOUËT (2006b) to quantify the effects of different trade strategies of the Mediterranean Partner Countries towards the EU. MIRAGE (Modelling International Relationships in Applied General Equilibrium) is a multi-market, multi-region recursive dynamic computable general equilibrium model for trade analysis developed by the CEPII (Centre d' Etudes Prospectives et d' Infor-

---

[16] This study has been also found as paper presented at the 10th Conference „Theories and Methods in Macroeconomics" Universite des Sciences Sociales, Toulouse, January 19-20 (FERABOLI, 2006).

mations Internationales). It is based on the assumption of imperfect competition for trade in industry and services and of perfect competition for agriculture. The horizontal product differentiation is linked to varieties and geographic origin and the vertical to one to two quality ranges depending on the country of origin of the product. A sequential dynamic set-up from 2006 to 2020 is applied that tracks factor productivity, including GDP expectations as given by the World Bank's development index. The dynamic nature of the model implies that an investment function is used to modify the stock of capital at each period. The model closes using unskilled labour imperfectly mobile between agricultural and non-agricultural activities. The SAM of GTAP version 6, aggregated into 24-regions and 16-commodities is used as database for the modelling exercise. Among the MPCs Morocco, Turkey and Tunisia are modelled separately (the other MPCs are included as the rest of North Africa and the rest of the Middle East). The studied commodities within the agricultural sector are rice, wheat, other cereal grains, fruit and vegetables, other agricultural products, sugar, plant-based fibres, meat and meat products and milk. The trade strategies of the MPCs are formulated within three simulation scenarios depending on the level of trade liberalisation (i.e. tariff dismantling) between the MPCs and the EU or the rest of the world. The first scenario simulates a free trade area between the EU and the MPCs, the second a south-north agreement where each of the MPCs separately forms a free trade area with the EU in industrial and agricultural products and the last one multilateral liberalisation. In the first scenario the growth in GDP is highest for North Africa (about 2 %) due to reduction in distortions and an increase of economic activity caused by higher exports to southern Mediterranean countries. Intersectoral reallocations of production are relatively smooth, with the exception of Turkey (with a decrease of about 30 % in rice production) and Tunisia (a decrease of about 38 % in wheat production). A north-south liberalisation results in a deterioration of the terms of trade for all MPCs, as imports are diverted from competitive producers (as Australia-New Zealand, Latin America) to less competitive ones (the EU). At the same time there are evident trade creation effects, especially in Morocco and Tunisia, where exports rise by about 42 and 46 % respectively. Social welfare only increases in Tunisia and Turkey (1.3 and 0.5 % respectively) and falls in all the other MPCs, by 0.1 to 0.9 %. The shifts in production are rather high, with for example a growth in the production of rice of about 904.7 % in Tunisia, 102.2 % in Turkey and 220.9 % in the rest of North Africa. Milk production falls in all MPCs from -0.3 % in the rest of the Middle East to -55.5 % in Morocco. Multilateral liberalisation provides the highest welfare gains for all MPCs (varying from 0.4 in the rest of the Middle East to 2.3 % in Tunisia). Due to trade creation effects the MPCs' export volumes expand by 11.4 %, compared with 8.8 % under the second scenario, and just 1.9 % under the first scenario. Reallocations of production factors for agricultural commodities are lower than in the previous scenario. Among the highest are a decrease of rice production in Morocco (-77.8 %) and in Turkey (-30.3 %), a decrease in wheat production in Tunisia, by about 65 % and a decrease in milk and meat production in Morocco, by about

40 % and 23 % respectively. The author concludes that the best trading strategy of the MPCs is that of multilateral liberalisation.

Static Partial equilibrium (PE) models

GRETHE (2003) developed a comparative-static, partial equilibrium model of the Turkish agricultural sector, named TURKSIM, to analyse the impacts of the customs union between Turkey and the EU[17]. The model assumes iso-elastic behavioural functions of farm supply on a regional level and demand. It covers 42 primary and processed agricultural commodities, which account for more than 86 % of the production value of Turkish agriculture. The author developed three scenarios to simulate i) the maintenance of the status quo, ii) full unilateral liberalisation and iii) agriculture being included in the customs union between Turkey and the EU. The base year for the simulation was the average of 1997-8 for the quantities of plant products and 1998-1999 for the quantities of animal products and the base year for prices was taken as the average over the period 1996-1998. The projections refer to 2006. Complete liberalisation of the agricultural sector would lead to significant static welfare gains of € 670 million. Including agricultural products in the customs union with the EU had very similar effects but lower terms of trade effects and the author treats this as a kind of political liberalisation. The distributional and allocative effects are significant. Liberalisation leads to a more equal distribution of real income, reduces intra-sectoral inequalities, but also reallocates resources from rural to urban areas.

AUGIER and GASIOREK (2003b) did a preliminary study, employing a partial equilibrium model of imperfect competition to examine the impacts of tariff reduction within the Barcelona Agreement on Morocco, focusing on the textile sector. The exogenous parameters of the model are based on detailed data at the firm level of the textile sector in Morocco. The first scenario involves an asymmetric reduction of import tariffs levied by Morocco of 50 % and results in a decline of the Moroccan textile and clothing sector, accompanied by a reduction of the production and exports from EU Mediterranean Member States. The results for Morocco are reversed when the costs to Moroccan firms in accessing the EU markets are reduced (second scenario). The same scenarios are run with lower aggregation of the data set for the Moroccan textile sector. Here an increase of market access to the EU clearly benefits exporting firms (second scenario) whereas in the first scenario exporting firms are completely eliminated from the Moroccan market and those that market domestically suffer a reduction in output of 63.9 %.

Of particular relevance is a study conducted within the EU-financed project EU-MED AgPol (BRITZ et al., 2006), where the partial equilibrium model CAPRI (Common Agricultural Policy Regionalised Impact) was used to simulate trade liberalisation be-

---

[17] Parts of this study have been published as GRETHE (2004) and GRETHE (2005).

tween the EU and the MPCs. The CAPRI modelling system was developed by the University of Bonn and is based on previous models, RAUMIS, SPEL/EU and WATSIM, developed by the same institution. It covers all the regions of the EU-25 disaggregated to sub-national regions at NUTS II level (Nomenclature des Unités Territoriales Statistiques) and is designed to simulate the EU's agricultural sector. The model is linked to the WATSIM trade system and thus employs partial equilibrium techniques to enable the modelling of global trade policies. In this application the Mediterranean region is aggregated to Turkey, Morocco and the rest of Mediterranean, while the EU-25 is disaggregated to the old EU-15 and the new EU-10. It covers the commodities typical to Mediterranean countries, such as fruits, vegetables and olive oil. The simulation followed similar lines to those used by KUIPER (2006) (see above), which allow the formulation of a partial bilateral liberalisation scenario, where import quotas are adjusted i.e. increased (to about twice the level of the base year) and Tariff Rate Quotas (TRQs) are expanded by about 50 %. The authors also ran a scenario in which, in addition to the bilateral liberalisation also simulates the G-20 proposals of the WTO negotiations, although they only provide limited information about the tariff cuts on the bound tariffs are integrated in the modelling framework. The results indicate that bilateral partial liberalisation would have hardly any impact on EU producers' prices and the quantities produced, but that it would have an impact on trade flows. For example under the full liberalisation scenario the imports of tomatoes from MPCs into the EU almost double, and imports of table olives and citrus fruits also increase substantially. Under full liberalisation there is an increase of Turkey's and Morocco's imports by about 60 % for each country and of about 350% for other Mediterranean countries. The EU's exports of cereals, and especially wheat, increase but there is also a significant growth in its imports of fruits and vegetables. The picture changes when the WTO negotiations are taken into account. In this case the producers' price in the EU decreases, particularly for livestock commodities. The authors explore the regional effects on farmers' incomes in the EU, focusing on regions in Spain and Italy. In both countries the income from tomato production increases across all NUTS II regions, but the income from citrus production declines, as a result of lower prices due to the substitution of the domestic production from imports. The authors expect welfare gains only to accrue to the largest northern EU Member States, while for other EU Member States the changes are not distinguishable.

Another static partial equilibrium model was developed by GARCIA ÁLVAREZ-COQUE et al. (2007) within the EU research project "Agricultural Trade Agreements" and was applied in the fresh tomato market. The model captures seasonal effects on the tomato market and thus enables the modelling of the entry price system for imports in the EU. The preliminary results of liberalisation scenarios on Moroccan exports into the EU indicate that the various liberalisation scenarios would affect negatively the EU market (reduction of EU sales by about 5 % which, in the event of the removal of the entry price system, could be up to 20 % in some months). The results reveal positive effects on Morocco especially when the liberalisation is multilateral and the tariff reductions apply to MFN suppliers. While being an interesting modelling exercise, the large numbers of parameters required mean that its extension to other markets would be a very time-consuming task. Market interactions though are not examined, and the elasticities used, are not calculated on a monthly level but on a yearly basis. Moreover the substitution elasticities are relatively high compared to those used in other modelling exercises (for example BRITZ et al., 2006; GRETHE, 2003; M'BAREK, 2002).

Spatial partial equilibrium model

M'BAREK (2002) developed a spatial equilibrium model to analyse the Euro-Med Agreements and their impact on Morocco and Tunisia. It is an interregional trade model with non-linear iso-elastic supply and demand functions, comparative static and synthetically calibrated to 1997 as the base year. It covers 19 primary agricultural commodities (soft wheat, hard wheat, barley, maize, soya cake, vegetable oils, olives, oranges, dates, tomatoes, early and main crop potatoes, sugar beet, sugar cane, beef, sheep meat, poultry meat, eggs and milk), seven supply regions (Tunisia, Morocco, EU, CEEC, USA, Argentina and the rest of the world) and six demand regions (Tunisia, Morocco, EU, CEEC, USA and the rest of the world). One interesting aspect of the study is that it introduces a land restriction, so that the production of a crop cannot exceed the available land for this specific crop. Transportation costs are given in the model either as costs per km or as fixed costs for a certain destination that were provided through interviews by freight companies. The simulations involve formulating a base run scenario and three main scenarios. The base run scenario is built for the year 2007 (implying that the projections are also for 2007) and includes the modalities for the agricultural sector according to the time schedule of the Doha WTO Round in 2003, the EU Eastward enlargement to include five countries (Estonia, Poland, Slovenia, the Czech Republic and Hungary), free trade for industrial products among the EU and the Maghreb countries and the formation of the Greater Arab Free Trade Area. The first main scenario simulates a partial and a full unilateral trade liberalisation between the EU, Morocco and Tunisia. A similar scheme is repeated in the case of bilateral and of multilateral trade liberalisation in the second and third main scenarios, respectively. The results indicate positive changes for Tunisia. Consumers enjoy benefits due lower prices, but at the expense of producers, particularly producers of cereals and livestock products. The production of cereals and animal commodities declines but the production of typical Mediterranean commodities increases to varying degrees under the different scenarios (for example tomato production increases by 901.3 % under full bilateral liberalisation and by 15.7 % under full unilateral liberalisation). In Morocco the changes in production, prices and demand are similar to those in Tunisia, but are more moderate. Producers again face

Empirical studies 89

lower prices and suffer welfare losses, while consumers benefit. Moving towards multilateral liberalisation the changes are more profound. The author also discusses the effects on agricultural employment, indicating that a reduction of production leads to lower employment in agriculture, which could cause further problems in rural development in the two countries. Although the study discusses the welfare gains for consumers and producers, the effects on taxpayers from reduced import tariff rents are not examined and thus conclusions for the overall social welfare or for the allocation of resources cannot be drawn.

Other models

Input-Output models

LORCA and VICÉNS (2000) report on the impacts of trade liberalisation within the framework of the Barcelona Agreement on economic growth in Egypt, Morocco, Turkey and Tunisia. Their model is based on Input-Output tables that utilise the macroeconomic characteristics and the economic development of these countries. By applying employment coefficients the model captures the effects of creating a free trade area among these countries in the period 2000-2004, i.e. after estimating econometrically the determinants of trade and the elasticities, it simulates a liberalisation scenario. Three simulation scenarios examine different options for liberalising trade between the EU and the four MPCs. Unilateral liberalisation from the EU leads to positive effects for all four countries on production (measured as the share of Added Value in GDP), exports and employment. Production will increase by between 0.5 % (in Tunisia) to 4.4 % (in Turkey) of GDP and generate more that 222,000 additional employment opportunities. The Mediterranean EU Member States will be most affected by the new competition, particular in comestible fruits, where the MPCs' exports to the EU will increase annually by 3.42 %. The second scenario simulates bilateral liberalisation leading to a decrease of about 30 to 40 % in prices for certain products in the MPCs (for example about -31 % and -40 % change in the prices of milk products and processed foods respectively), leaving the producers in these countries worse off and the consumers better off. This scenario not only models a complete dismantling of tariffs but also a reduction of the European domestic support, as foreseen by the WTO negotiations (i.e. the reduction of 55 % of internal subsidies affecting trade).The third scenario involves an asymmetrical reciprocal liberalisation together with development of a structural tool with MPCs adopting a European Guidance and Guarantee Fund (EAGGF) orientation. The results reveal gains for innovators in agriculture in the MPCs and losses for Mediterranean EU producers due to high competition. Rent-seeking behaviour could also occur in the MPCs depending on how efficiently the EAFFG oriented support for structural changes is distributed. The results of this modelling exercise are also included in a report under the framework of the FEMISE network (RADWAN et al., 2003) on the impacts of agricultural trade liberalisation in the context of the Euro-Med Agreements.

Mathematical programming models

A regional agricultural sectoral model called TASM-EU (Turkish Agricultural Simulation Model) was employed by CAKMAK and KASNAKOGLU (2003) to quantify the effects of Turkey's membership in the EU on its agricultural sector. It is a non-linear optimisation model which maximises consumer and producer surplus. It describes total national supply and use, with production being decomposed into sub-models for Turkey's four main regions. Consumer behaviour is considered as price dependent, meaning that market clearing commodity prices are endogenous. It is calibrated using positive mathematical programming techniques. The model contains more than 200 activities that describe the production of about 50 commodities. The base period for the model is the average of the years 1997-1999 and it is solved for 2005. The application runs four simulation scenarios. The first assumes that Turkey will not become a member of the EU and only adjusts population and income growth to 2005, using estimates from the Turkish statistical office and price projections from FAPRI. The next three scenarios examine various options of Turkey's membership of the EU. Membership of the EU would lead to welfare gains for consumers (increasing consumer surplus by about 12 %) but to welfare losses for producers (decreasing producer surplus by about 16 %). The prices (both producers and consumers) in Turkey are adjusted to EU ones, resulting in lower prices both for producers and consumers. The volume and the value of livestock production decrease by about 22 % and 40 % in the EU membership scenarios, reflecting the backward production conditions in Turkey. By contrast crop values and volumes increase. For example, the production of durum wheat expands by 4 %, that of cotton and of sugar beet by about 5 and 4 % respectively, and fruit and vegetable production goes up by about the same level. Only the production of soft wheat declines by about 10 %. Exports of crop products decline by about 30 % while imports of livestock products increase four fold. There are no changes in the trade status of Turkey compared to the base run (and it remains a net exporter of crops and a net importer of livestock products). It should be noted that when compensatory payments are introduced in Turkey due to EU membership, they further compensate to the decline in production, due to set-aside requirements. The level of consumption increases because of a fall in consumer prices, by about 40 % which leads to consumer expenditure being about 25 % lower.

4.2.3 Outlook on empirical studies and identification of further research needs

By taking a deeper look at the existing empirical studies it can be concluded that, although agricultural policies affecting the Mediterranean basin have been empirically discussed, the majority of studies focus on the impacts of trade liberalisation between the EU and single MPCs within the Euro-Med Agreements in terms of the entire economy of the MPCs examined. Fewer studies analyse the effects of the reformed CAP on EU Member States and MPCs and only the study by BRITZ et al. (2006) reveals simultaneous effects brought about by the new agricultural policy scheme in the EU and the conclusion of the Euro-Med Agreements.

In detail, the scenarios analysed in these studies are related to tariff cuts between the examined Mediterranean country(ies) and the EU. Because the opening of the EU markets to the MPCs under the Euro-Med Agreements is a stepwise procedure, almost all the studies simulate scenarios that examine various degrees of tariff reduction (usually 50 and 100 %). This is true of AUGIER and GASIOREK (2003a and b), HARRISON et al. (1997), HOEKMAN et al. (2001), HOSOE (2001), MERCENIER and YELDAN (1997), RUTHERFORD et al. (1997) and the studies that use dynamic CGE models, whereas liberalisation can be either unilateral (from the side of the MPCs) or bilateral. KUIPER (2006) followed a different scheme, formulating a base assumption, where the policy variables are adjusted (shocked) and thereby approximate the policy framework in the year that the results refer to. These are then simulated into a full bilateral liberalisation between the examined MPCs and the EU, thus providing the lower and upper bounds of the forthcoming liberalisation. A similar scheme is followed by BRITZ et al. (2006). Additional policies are simulated in some cases, such as adjustments to fiscal policies (for example KONAN and MASKUS, 2000), while in all the studies except those of KUIPER (2006) and BRITZ et al. (2006) the EU agricultural policy is not simulated. It should be noted that the nature of the Euro-Med Agreements makes it particularly difficult to formulate scenarios. A number of countries are negotiating separately with the EU through association agreements, and thus the implementation of the agreements varies and is at different stages. In addition the preferences granted by the EU to one country depend on those granted to other countries (GARCIA ÁLRVAREZ-COQUE et al., 2007).

The results of these exercises mainly focus on the whole economy of the non-EU countries. The authors generally agree that liberalisation will result in welfare gains for the EU, increase of its exports to non-EU Mediterranean countries and higher producer prices in the MPCs. The magnitude of the effects varies according to the importance of the liberalised sectors for the EU markets (for example liberalisation in manufactures or/and services). KUIPER (2006) examines regional effects on the EU and finds that the northern EU countries would benefit from improved market access to North African countries for cereals, animal and dairy products, while the Mediterranean EU countries would face increased imports of vegetable oils.

It is worth noting that the results of the CGE models depend on the assumption of capital mobility. Nevertheless, as GARCIA ÁLVAREZ-COQUE et al. (2006) argue, the impacts of trade liberalisation are often sectorally and regionally specific. Very often, and at least in the short term, it is difficult to find alternative opportunities for labour and capital. This is especially relevant in the case of rural areas of MPCs, where agriculture is very often the sole employment source. In such circumstances the assumption of mobile production factors needs to be checked for its plausibility. It should also be noted that the use of the equivalent variation in the GTAP model to measure changes in income does not provide insights on the distribution of welfare changes among different groups (producers, consumers, taxpayers).

With the exception of the studies of KUIPER (2006) and BRITZ et al. (2006) all the other studies give to the agricultural sector a limited role and either represented it in an aggregated way or through a limited number of commodities. This high level of aggregation does not allow drawing conclusions on the agricultural sector. Changes in the main commodities within the region are not captured by the modelling exercises, thus undermining the plausibility of the empirical studies. It should be also noted that in most of the CGE models liberalisation is focused on manufacture goods and services and not on agricultural commodities.

There are further agriculturally related issues that should be taken into account. The EU enlargement of twelve new Member States has expanded the EU domestic market, which could help narrow the existing opposing positions within the EU for liberalising agricultural trade between it and MPCs (GARCIA-ALVAREZ-COQUE, 2002). The CAP reform creates a new framework for European farmers and, depending on how the reform is implemented, this could break the north-south conflict of interests among European farmers (GARCIA-ALVAREZ-COQUE, 2002).

The above studies cannot answer the question of the extent to which Mediterranean agriculture will be affected by the forthcoming policy changes or by the recently concluded agricultural policy reforms. Further empirical analysis is required which employs adjusted modelling tools. The need to base future negotiations on sound empirical findings is evident in the growing interest of the EU and those countries involved in the Agreements to finance research projects and create research networks. It should be noted that most research findings to date have been the outcomes of EU funded research projects.

Empirical analysis of Mediterranean agriculture and the impacts of trade policy reforms on it would appear to require a multi-commodity and multi-region equilibrium model. This should be focused on the agricultural sector and on commodities that are typical for the region. It would be preferable to model these commodities separately and in a more disaggregated form, i.e. it should focus on specific fruits and vegetables rather than viewing these as one aggregated commodity. To capture the interactions between the EU and the MPCs it is necessary to include both blocks of countries in the regional aggregation scheme, and it would be interesting to break the EU down into southern and northern Member States and the MPCs into single countries. In this way the different regional effects on the EU and on each MPC could be shown better, thus avoiding problems of generalising the effects in the EU and the MPCs. Simulation scenarios could interestingly be used to see the extent to which changes in European agricultural policy affect MPC countries that are joined with the EU through a preferential system and the extent to which any future openness of EU markets to the MPCs would affect the EU Mediterranean Member States.

A partial equilibrium model at this level of regional aggregation, covering the main agricultural commodities could provide appropriate insights into the impacts of policy changes on Mediterranean agriculture. Due to practical difficulties in finding the exogenous parameters, a model employing a straightforward equation system might prove to be sufficient for such an analysis.

4.3 Overview of AGRISIM

AGRISIM (Agricultural Simulations Model) is a partial-equilibrium, multi-commodity, multi-region model. It is comparatively static in nature, deterministic and has non linear isoelastic supply and demand functions (PUSTOVIT, 2003; SCHMITZ, 2002). Trade is modelled as net trade. Policy interventions are generally considered as changes in nominal protection rates, price transmission elasticities, minimum producer prices, production quotas and subsidies. Through shift coefficients in the demand and supply functions additional variables can be simulated, such as population and income growth. The base version of the model includes nine commodities: wheat, coarse grains, rice, oilseeds, sugar, milk, beef, pork and poultry meat. The database was recently updated up to the year 2001 and was extended, as will be described later.

4.3.1 Model description

The main structure of the model is shown in Figure 4.1, following the suggestions of RONINGEN (1997) describing a multi-market multi-region partial equilibrium model and the main structure of the model SWOPSIM (Static World Policy Simulation Modelling Framework (RONINGEN et al., 1991). The regions are connected with each other through a market clearing mechanism, and the world market price that results from this mechanism is fed into domestic markets through domestic prices. The net trade summed from all regions, arrived at by the difference between supply and demand, is fed back into the world market clearing mechanism.

Extending the data base of the model and modifying its structure allow further analysis of trade policies by changing the level of applied tariffs and export subsidies on a multilateral level for all the countries in the model.

The results involve changes in prices (domestic and world market prices), produced quantities, consumption, net trade and welfare

Figure 4.1: Simulations-routine in AGRISIM; example of 2 markets – 2 commodities

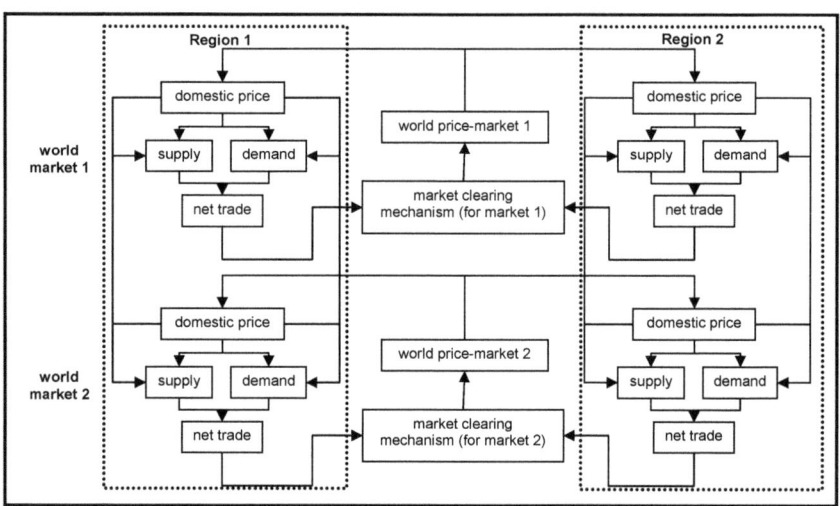

Source: Own illustration based on RONINGEN (1997)

The rest of this section introduces the main equations of the model in detail. The sets of the model are defined as:

r = all regions;

i,j = all markets

Volumes

Supply function

The quantity supplied is given by an iso-elastic function. Cross price effects between the markets are considered through cross price elasticities. The price that influences the supply is the producer incentive price.

$$S_{i,r} = s_{i,r} \cdot \prod_j \left(p_{i,r}^P\right)^{\varepsilon_{i,j,r}^S} \cdot SHIFT_{i,r}^S \tag{1a}$$

with:

$S_{i,r}$ = Domestic supply of product i in region r

$s_{i,r}$ = Calibration parameter of supply function

$p^P_{i,r}$ = Producer incentive price

$\varepsilon^S_{i,j,r}$ = Own and cross price elasticity of supply

$SHIFT^S_{i,r}$ = Supply shifter (yield and other shifts)

This equation is slightly deferentiated when binding production quotas are applied in a commodity market. In this case the price that influences the supply is the quota equivalent price, as shown in equation 1b.

$$S_{i,r} = s_{i,r} \cdot \prod_j \left(p^P_{i,r} - p^{Quo}_{j,r}\right)^{\varepsilon^S_{i,j,r}} \cdot SHIFT^S_{i,r} \qquad (1b)$$

with:

$p^{Quo}_{j,r}$ = Difference between producer incentive price and quota equivalent price

Yield function

The yield in the model is given by the following equation. It is calculated according to the data parameters of the base year and remains constant unless the shift factors of technical progress or annual yield growth are modelled. Using the price elasticity of yield allows the possibility of differentiating the effects of price on the yield. Due to lack of data this possibility is not used within the current simulations.

$$Y_{i,r} = y_{i,r} \cdot \left(p^P_{i,r}\right)^{\varepsilon^Y_{i,j,r}} \cdot SHIFT^Y_{i,r} \qquad (2)$$

with:

$Y_{i,r}$ = Yield of product i in region r

$y_{i,r}$ = Calibration parameter of yield function

$\varepsilon^Y_{i,j,r}$ = Price elasticity of yield with respect to own price

$\Delta^Y_{i,r}$ = Yield shifter (e.g. annual yield growth trend, technical progress)

Area function

The acreage (or number of animals for livestock commodities) is calculated by dividing the overall supply of a commodity by the yield:

$$A_{i,r} = \frac{S_{i,r}}{Y_{i,r}} \tag{3}$$

where:

$A_{i,r}$ = Area (or number of animals) of product i in region r

Seed demand

The demand for seeds is determined by multiplying domestic supply with a calibration parameter of the seed demand function and a shifter, which enables modelling factors that could result in a shift of the function such as, for example, technical progress.

$$D_{i,r}^{S} = d_{i,r}^{S} \cdot S_{i,r} \cdot SHIFT_{i,r}^{SD} \tag{4}$$

with:

$D_{i,r}^{S}$ = Seed demand of product i in region r

$d_{i,r}^{S}$ = Calibration parameter of seed demand function

$SHIFT_{i,r}^{SD}$ = Seed demand shifter (e.g. technical progress)

Feed demand

The feed demand is given by an iso-elastic function. Due to the lack of data the elasticities are identical to the elasticities used in equation 6. Through the shifter it is possible to model changes in animal numbers.

$$D_{i,r}^{F} = d_{i,r}^{F} \cdot \prod_{j}\left(p_{j,r}^{C}\right)^{e_{i,j,r}^{DF}} \cdot SHIFT_{i,r}^{FD} \tag{5}$$

with:

$D_{i,r}^{F}$ = Feed demand of product i in region r

$d_{i,r}^{F}$ = Calibration parameter of feed demand function

$p_{j,r}^{C}$ = Consumption price

Empirical studies

$\varepsilon_{i,j,r}^{DF}$ = Own- and cross-price elasticity of feed demand

$SHIFT_{i,r}^{FD}$ = Feed demand shifter (e.g. changes of animal numbers)

Non-agricultural demand (food demand)

Non agricultural demand, or food demand, is determined in a similar way to feed demand. Although the model is static in nature, it is possible to use the shifter of this function to consider dynamic effects, such as changes in income or population.

$$D_{i,r}^{NA} = d_{i,r}^{NA} \cdot \prod_{j} (p_{j,r}^{C})^{\varepsilon_{i,j,r}^{NA}} \cdot SHIFT_{i,r}^{NA} \qquad (6)$$

with:

$D_{i,r}^{NA}$ = Domestic non agricultural demand of product i in region r

$d_{i,r}^{NA}$ = Calibration parameter of domestic non agricultural demand function

$\varepsilon_{i,j,r}^{NA}$ = Own- and cross-price elasticity of non-agricultural demand

$SHIFT_{i,r}^{NA}$ = Non agricultural demand shifter (e.g. change in income, population)

Waste

As with the seed demand, waste is dependent on the quantities supplied, a shifter for modelling technical progress and a calibration parameter.

$$W_{i,r} = w_{i,r} \cdot S_{i,r} \cdot SHIFT_{i,r}^{W} \qquad (7)$$

with:

$W_{i,r}$ = Waste of product i in region r

$w_{i,r}$ = Calibration parameter of waste function

$SHIFT_{i,r}^{W}$ = Waste shifter (e.g. technical progress)

Net trade

Net trade is calculated as the difference between the quantities supplied, the stock and all the components of demand i.e. seed demand, feed demand, non agricultural demand and waste.

$$NT_{i,r} = S_{i,r} + ST_{i,r}^{BY} - D_{i,r}^{S} - D_{i,r}^{F} - D_{i,r}^{NA} - W_{i,r} - Adj_{i,r} \qquad (8)$$

with:

$NT_{i,r}$ = Net trade of product s in region r

$ST_{i,r}^{BY}$ = Change in stocks of product s in region r in base year (constant)

$Adj_{i,r}$ = statistical adjustments

Prices

There are four price definitions in the basic version model, namely border price, domestic price, producer incentive price and consumer price.

Border price

The border price is defined in relation to a reference price. The USA is used as the reference country in the model and thus USA border prices are the reference border prices. A region's border price for a certain commodity is therefore determined by the USA border price and the difference between the border price of the region and the reference border price in the base year.

$$p_{i,r}^{B} = p_{i,ref}^{B} + \left(p_{i,r}^{BY} - p_{i,ref}^{BY} \right) \qquad (9)$$

with:

$p_{i,r}^{B}$ = Border price of product i in region r

$p_{i,ref}^{B}$ = Reference border price of product i (USA border price)

$p_{i,r}^{BY}$ = Border price of product i in region r in base year

$p_{i,ref}^{BY}$ = Reference border price of product i in base year

Domestic price

The domestic price is determined by the nominal protection coefficient i.e. the relationship between border and domestic prices and the price reaction to the border prices. The price transmission elasticity gives the level of the relationship between the border and the domestic price. When $\varepsilon_{i,r}^{p} = 0$, then changes in the world market price (and thus of the border price) do not affect domestic prices and when $\varepsilon_{i,r}^{p} = 1$ then changes in world market prices are transmitted fully to the domestic market.

Empirical studies

Various trade policies can be simulated, depending on the level of the nominal protection coefficient and the price transmission elasticity. In this study it is assumed that $\varepsilon^P_{i,r} = 1$.

$$p^D_{i,r} = NPC_{i,r} \cdot \left(p^B_{i,r}\right)^{\varepsilon^P_{i,r}} \tag{10}$$

with:

$p^D_{i,r}$ = Domestic price of product i in region r

$NPC_{i,r}$ = Nominal protection coefficient

$\varepsilon^P_{i,r}$ = Price transmission elasticity

Producer incentive price

The producer incentive price is calculated endogenously and is equal to the domestic price and the element of subsidies that influence production, as given in equation (11). The effects of decoupling are modelled through the "production effectiveness" coefficient, showing how much the introduction of decoupled payments acts as an incentive for production and thus influences the quantity produced and the prices that farmers actually receive.

$$p^P_{i,r} = p^D_{i,r} + \sum_{Sub} \alpha_{Sub} Z_{Sub} \tag{11}$$

with:

α_{Sub} = Production-effectiveness

Z_{Sub} = Subsidy per ton

Consumer price

Due to lack of data the consumer price used in this study is considered to be the same as the domestic price. Theoretically, and if the data exist, it is possible to add retail margins as a further factor affecting the consumer price.

$$p^C_{i,r} = p^D_{i,r} + RM_{i,r} \tag{12}$$

with:

$p^C_{i,r}$ = Consumer price

$RM_{i,r}$ = Retail margin

Market Clearing

The equilibrium conditions are given in equations 13 and 14. The market is cleared when the sum of the net trade for all regions and for all commodities is equal to zero.

$$\sum_r NT_{i,r} = 0 \tag{13}$$

$$\sum_i \sum_r NT_{i,r} = 0 \tag{14}$$

Welfare

Welfare effects are measured using the surplus concept. Welfare changes consist of changes in consumer surplus, in producer surplus, in the quota owner surplus (this is relevant only for those markets, where a binding production quota is applied) and the effect on taxpayers or the budget.

Consumer surplus

The effects on the consumer side follow the Marshallian concept of using willingness to pay as measurement for consumer surplus. The consumer surplus is the difference between what the consumer is ready to pay (willingness to pay) over what he actually pays (real expenditure).

For example the change of the willingness to pay in a single market because of an increase of the demand from $D_{i,1}^{NA}$ to $D_{i,2}^{NA}$ is illustrated as the integral of the respective interval of the inverse demand curve $P_i(D_i^{NA})$:

$$\Delta Z_i = \int_{D_{i,1}^{NA}}^{D_{i,2}^{NA}} P_i(D_i^{NA}) dD_i^{NA} \tag{15}$$

with:

ΔZ_i = change of the willingness to pay

Therefore the change of the consumer surplus in a single market is given by the following equation:

$$\Delta CS_i = \Delta Z_i - P_{i2}^C D_i^{NA} + P_{i1}^C D_i^{NA} \tag{16}$$

with:

ΔCS_i = change of the consumer surplus

Empirical studies

The consumer surplus can be alternatively given direct from the integral of the demand function. Therefore the change of the consumer rent due to an increase of the consumer prices from $P_{i,1}^C$ to $P_{i,2}^C$ would be:

$$\Delta CS_i = - \int_{P_{i,1}^C}^{P_{i,2}^C} D_i^{NA}(P_i^C) dP_i^C \tag{17}$$

In the model the changes of the consumer surplus of a region r and a market (commodity) i in a simulated scenario expressed as deviation from the Base Run Assumption (BA) is given by the following equation:

$$CS_{i,scba} = -1 D_{i,scba}^{NA} \frac{1}{1+\varepsilon_i^{NA}} \left(\frac{P_{i,scba}^{CI\ (1+\varepsilon_i^{NA})} - P_{i,BA}^{CI\ (1+\varepsilon_i^{NA})}}{1000} \right) \tag{18}$$

where $P_{i,j,r}^{CI}$ is the income normalised consumer price.

Due to path-dependency problems in the case of multiple price changes (for more see JUST et al., 2004, pp.102-105), both integration paths are calculated in the model. The change in consumer surplus is the arithmetic average of the two paths.

The changes of the consumer surplus in all markets are given by the sum of the changes of the consumer surplus in the single markets:

$$CS_{r,scba} = \sum_i CS_i \tag{19}$$

The feed and seed consumer rents are computed in an analogous manner.

Producer surplus

The welfare effects for producers are captured following the Marshallian concept of quasi-rent, which is equivalent to the producer surplus and is given as the excess of gross receipts over total variable costs. In a single market the increase of production costs because of an increase of supply from $S_{i,1}$ to $S_{i,2}$ are given by the integral of the respective interval of the marginal cost curve $P_i^P(S_i)$:

$$\Delta K_i = \int_{S_{i,1}}^{S_{i,2}} P_i^P(S_i) dS_i \tag{20}$$

with:

ΔK_i = change of total production costs

The change of the producer surplus in a single market is thus:

$$\Delta PS_i = P_{i,2}^P S_{i,2} - P_{i,1}^P S_{i,1} - \Delta K_i \tag{21}$$

with:

ΔPS_i = change of the producer surplus

Alternatively the welfare effects on the producer side can be given direct by the integral of the supply curve. Therefore for an increase of the price from $P_{i,1}^P$ to $P_{i,2}^P$ the change of the producer surplus is:

$$\Delta PS_i = -\int_{P_{i,1}^P}^{P_{i,2}^P} S_i(P_i^P) dP_i^P \tag{22}$$

In the model the resultant producer surplus in a single market for a simulated scenario (scba) caused by a change in producers' incentive price is expressed as deviation from the base run (BA) is given by the equation:

$$PS_{i,scba} = S_i \frac{1}{1+\varepsilon_i^S} \frac{(P_{i,scba}^P - P_{j,scba}^{Quo})^{1+\varepsilon_i^S} - (P_{i,BA}^P - P_{j,BA}^{Quo})^{1+\varepsilon_i^S}}{1000} \tag{23}$$

Binding production quotas have a negative impact on producer surplus (the rent of the quota owner is calculated separately). Since the model separately includes seed and feed demand, the effects of price changes on these markets are explicitly examined and the resulting welfare effects on the supply side are calculated in an analogous fashion to changes in the consumer surplus and added to the producer surplus.

The total changes in the producer surplus are given by the sum of the changes on single markets:

$$PS_{r,scba} = \sum PS_{i,scba} \tag{24}$$

Empirical studies

Rent for the quota owners

In the event of binding production quotas, a rent for the quota owners entails. The quota rent is calculated by the price difference between the producer incentive price and the quota equivalent price multiplied by the binding production quota. The binding production quota is given exogenous.

$$QR_{j,r} = p_{j,r}^{Quo} \cdot QUOTA_{j,r} \tag{25}$$

with:

$QR_{j,r}$ = quota rent

$QUOTA_{j,r}$ = binding production quota

In the model the resultant quota owner surplus in a single market and for a simulated scenario (scba) is expressed as deviation from the Base Run (BA) (equation 26).

$$QR_{i,scba} = \frac{(P_{j,scba}^{Quo} \cdot QUOTA_{j,scba}) - (P_{j,BA}^{Quo} \cdot QUOTA_{j,BA})}{1000} \tag{26}$$

The total changes in the rent for the quota owners are given again by the sum of the changes over the single markets:

$$QR_{r,scba} = \sum QR_{i,scba} \tag{27}$$

Taxpayers' effect

The budget expenditure or taxpayers' expenditure or revenue consists of two main elements. The first is the change in expenditure on custom duties, such as export subsidies, and/or the revenue from custom duties, such as import levies. The second is related to the budget expenditure for financing domestic policy measures and measures the change in expenditure for direct, input and general subsidies. A correction is made for EU Member States due to the intra-community financing system of the EU. This correction is based on the national contribution that each Member State makes to the EU budget, the resources of the EU paid or collected by each Member State, (i.e. export subsidies are financed by the EU, whereas the revenue from import tariffs is collected by Member States and paid into the EU budget) and on the common financing of the agricultural policy measures, which also occur at the level of the Member States (i.e. direct, input and general subsidies for agriculture are financed by the EU budget and not from resources of individual Member States).

4.3.2 Calibration of Elasticities

The initial price elasticities taken from the literature violate the theoretical conditions of microeconomics and thus, must be adjusted before being used in the model.

On the supply side the conditions of homogeneity and symmetry need to be taken into account.

The homogeneity condition is shown in equation 28. The supply function is homogenous of degree zero when the supplied quantity remains stable after a change (increase) in the commodity prices by the same percent (KIRSCHKE and JECHLITSCHKA, 2002, pp. 165 – 168).

Derivation and manipulation of the supply function gives:

$$\varepsilon_{S_1,p_1} + \varepsilon_{S_1,p_2} + \ldots + \varepsilon_{S_1,p_i} + \ldots + \varepsilon_{S_1,p_n} = 0 \tag{28}$$

with:

ε_{S_1,p_1} = price elasticity of supply

p_i = price for product i

for i = 1 to n

The symmetry condition of the supply side is expressed in equation 29. According to this changes in the quantity of one product caused by changes in the price of another product are equal to the changes in the quantity of the second product due to changes in the price of the first product (KIRSCHKE and JECHLITSCHKA, 2002, p.167).

$$\varepsilon_{S_1,p_2} p_1 S_1 = \varepsilon_{S_2,p_1} p_2 S_2 \tag{29}$$

with:

$S_{1,2}$ = Supplied quantities for products 1, 2

$p_{1,2}$ = price for products 1,2

ε_{S_i,p_j} = cross price elasticity of supply,

for i, j = 1,2 and i ≠ j

The same approach is used to express homogeneity and the symmetry condition for the demand function and the price elasticities of demand respectively (NICHOLSON, 1995, p. 209 and KIRSCHKE and JECHLITSCHKA, 2002, pp. 171-174). Since income elasticities are not implemented in this version of the model, the adding-up condition is not discussed.

4.3.3 Database

The updating and extension of the comprehensive AGRISIM database has been one of the key aspects of this project and one of the most time intensive tasks. The database contains raw information about primary and processed commodities and feeds the model with all the necessary exogenous parameters.

The model covers the whole world, aggregated into 17 regions of which, depending on the focus of the simulations, 56 can be modelled separately. It also covers 29 commodities, which are aggregated into commodity markets (for example all oilseeds are aggregated together), again depending on the focus of the analysis to be carried out.

An overview of the commodities and countries covered by AGRISIM is given in Table 4.2.

Table 4.2: Commodities' and countries' list

Commodities	Apples*	Rice	Pig meat
	Coarse grains (barley, maize, millet, oats, rye, sorghum, triticale, other cereals)	Oilseeds (rape and mustard seed, soya beans, sunflower seed)	Poultry meat (chicken, duck, goose, turkey meat, other poultry)
	Beef	Sugar	Tobacco*
	Cotton*	Olive oil*	Tomatoes*
	Milk	Oranges*	Wheat
Countries	Australia	Iceland	Norway
	Algeria*	India	Poland
	Brazil	Israel*	Romania
	Belarus	Japan	Russian Federation
	Bulgaria	Jordan*	Slovakia
	Canada	Korea, Republic of	Slovenia
	China	Latvia	South Africa
	Cuba	Lebanon*	Switzerland
	Cyprus*	Libya*	Syria*
	Czech Republic	Lithuania	Thailand
	Egypt*	Malta*	Tunisia*
	Estonia	Mexico	Turkey
	EU-15 (data for each country)	Morocco*	Ukraine
		New Zealand	USA
	Hungary	Norway	Rest of World

Notes: * New commodities and countries added in AGRISIM
Source: AGRISIM database

The database consists of three main parts, data on volumes and prices, data on trade policies and the elasticities.

Time series data from 1975 to 2001 of the volumes of production, commodity balances and population are derived from FAOSTAT.

Time series from 1986 to 2001 containing information on trade policies are taken from the PSE and CSE database of the OECD (2006). Other supplementary sources are used for those counties and/or commodities that are not included in the PSE databases. Ad-valorem applied tariffs are derived from AMAD, TARIC and the Market Access Database of the EU. The same sources are used for any existing tariff rate quotas, which are first converted to ad-valorem equivalents and then fed into the model. Export subsidies for 1995 to 2004 are taken from the WTO secretariat. More specifically, the tariffs applied to the Mediterranean commodities of all countries in the model and all commodities from the Mediterranean countries are taken from TRAINS.

The elasticities are derived from three main sources. Initially they were all taken from the database of the SWOPSIM model developed on behalf of the United States Department of Agriculture (USDA) (SULLIVAN et al., 1992). Later these were supplemented by data for Central and Eastern European Countries taken from the database of the CEEC-ASIM model developed on behalf of the IAMO (Leibniz Institute of Agricultural Development in Central and Eastern Europe) (WAHL et al., 2000). The recent updates and extensions of the model have drawn on additional sources, including the database of FAPRI and the USDA (SEALE et al., 2003). For Turkey and Morocco and for Tunisia the elasticities for tomatoes, oranges, apples and olive oil are taken from GRETHE (2003) and M´BAREK (2002) respectively.

The use of various data sources creates the need to check the consistency of data before using them in the model, due to the different definitions of commodities that are used.

It is worth noting that a special process is followed for modelling some particular commodities, such as sugar, vegetable oil, olive oil and cotton. These are the products derived from the processing of raw commodities, such as sugar beet and cane, oilseeds, olives and seed cotton. Here the data on the volumes of trade, the quantities consumed, the elasticities and policy refer to the processed commodities, whereas data on the produced quantities and farm gate prices refer to the raw commodities. In these cases the raw products are balanced with the processed ones by using empirical conversion factors, with the calculated equivalent quantities and prices being used in the model.

Finally, for milk in the EU regions, the term $p_{j,r}^{Quo}$ of equaltion 1b has been derived from GRAMS (2004, pp. 88-90 and p. 229).

Empirical studies

4.3.4 Technical Issues

The model is written in GAMS and uses the solver CONOPT 2.

Figure 4.2 shows the structure of the model from a technical point of view. The model consists of three parts: the preparation of data, the calibration of elasticities and the core of the model, where the simulations are run.

Figure 4.2: The modules of AGRISIM and their inter-connections

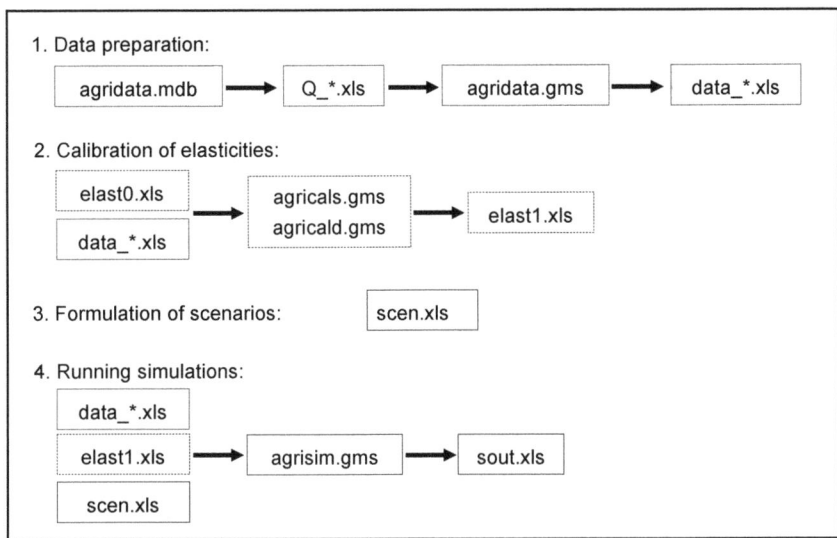

Source: Own illustration

The raw data are stored in a Microsoft Access file named „agridata.mdb". This can be searched for the relevant data for the simulations to be run, which are then exported to EXCEL files (named as Q_*.xls). These files are read and, if needed, further processed by the "agridata.gms" routine. This process ensures that the data are consistent for every single country and market in the database. The processed data, aggregated into "products" and "regions" are re-exported to EXCEL files, (now named "data_*.xls"). Before any processing of the raw data takes place, i.e. before the creation of the Q_*.xls files, the data are checked for their consistency. For example checks are made as to whether the export and import quotations of trade flows match and if data from the different data sources refer to the same product i.e. that they use the same product definitions.

The second step is calibrating the elasticities. The initial elasticities derived from SWOPSIM and other sources are contained in the file "elast0.xls" Through the routines "agricals.gms" and "agricald.gms" and by feeding into them the processed data containing information for each region ("data_*.xls") the calibrated elasticities are exported to the file "elast1.xls"

After the above steps the model is ready to run. The routine "agrisim.gms" is fed with the prepared data, the calibrated elasticities and the formulated scenarios (file "scen.xls") and the results exported to the file "sout.xls".

4.3.5 Limitations of the model

The forthcoming change in the agricultural policy regime of Mediterranean countries, particularly the creation of a Free Trade Area between the EU and the MPCs, combined with a lack of empirical studies creates the need to analyse empirically their impacts. AGRISIM was extended and adjusted to perform the analysis.

The extensions include an update of the data base of the model, to cover new commodities and build a different regional composition. The commodities selected are typical for the Mediterranean region and also important part of the external trade of Mediterranean countries, such as olive oil, tomatoes, apples and oranges. To capture the precise welfare effects on individual EU Member States, the calculation of the budget effects has been corrected in line with the EU intra-community financing system.

Even though great care and attention has been given to transforming the model, there are still certain limitations that need to be taken into account when interpreting the results.

The model is comparative-static in nature and, although shift factors and the possibility of modelling population growth allow some dynamic aspects to be captured, the results must be seen as comparative-static. For example non-trend changes in prices and quantities or in the behaviour of consumers and producers observed in reality can only be produced in the model by assumptions. Thus, the model is not suited for forecasts and the results should be interpreted as possible trends rather than as precise figures. However, the model is suitable for comparing what happens under different policy scenarios, depending on how the simulations are formulated. Additionally, because of its comparative-static nature the model underestimates the true gains from trade and from liberalisation, since in reality these gains are commonly much higher than the model results. Trade liberalisation is not a static procedure, but a dynamic one, which offers opportunities and can act as a catalyst for changes in the structure of production, the evolution of employment, and the organisation of supply chains.

The second limitation has to do with the exogenous parameters of the model. Several data sources have been used, which are not always consistent with each other. Although it is quite easy to obtain time series of quantitative data, it is very difficult to find reliable time series data of domestic or world market prices, thus making it necessary to use different data sources. It is assumed that the domestic prices are determined by a reference world market price, applied tariffs and export subsidies. Nevertheless, there can be other barriers to trade, such as non tariff barriers and negative protection, which are difficult to quantify and to measure and are not taken into consideration.

A further problem has to do with the elasticities used in the model. Although they are calibrated under the assumptions of microeconomic theory, questions can be asked about the reliability of the initial elasticities. These are derived from various secondary sources as described above (chapter 4.3.3). Ideally it would be to estimate econometrically all the initial elasticities. However, such an estimation would require reliable time series of domestic prices for all the countries in the world, which do not exist. Constructing such a database would be a very time- and effort-intensive task, far beyond the scope of this study. A second best option to overcome this problem is to carry out a sensitivity analysis based on the initial elasticities. The simulations can be run using different values of the initial elasticities, thus allowing a testing of the effects of the elasticities on the results.

Overall these limitations do not undermine the value of AGRISIM as a tool for empirically analysing the effects of the altered policy regime for Mediterranean countries. Compared to other partial equilibrium multi-commodity and multi-region models, it covers typical commodities for the Mediterranean region and countries at a non-aggregated level and through the new regional composition of the model, the regional effects on the EU-Member States can be captured.

What differentiates this study from previous empirical analyses, described earlier in this chapter is the commodity and regional composition of the model and its partial equilibrium nature, which allows a more detailed analysis of the agricultural sector. The model considers the main commodities that are traded around the Mediterranean basin capturing north-south and south-north trade flows. Due to the complexity of the market in fruit and vegetables and differences in trade policy that are applied not only to specific products but also to according to season, the model focuses on a few key products (apples, oranges and tomatoes) which are the most important for trade between the EU and the MPCs. Additionally, all Mediterranean Countries, both EU Member States and MPCs are modelled. By breaking the EU down into the Mediterranean Member States (Greece, Italy and Spain), the rest of the EU-15 and new Member States the model is appropriate for revealing regional disparities around the Mediterranean and within the EU.

5 Simulations with AGRISIM and model results

For the purposes of this study and in order to keep the model manageable the commodities and the countries that AGRISIM includes have been aggregated together. Table 5.1 shows the regional aggregation and commodity composition used in the simulations.

Table 5.1: Levels of aggregation used in the AGRISIM Database

Regions		Products	
GRE	Greece	APPL	Apples
ITA	Italy	ORAN	Oranges
ESP	Spain	TOMA	Tomatoes
E12	Rest of EU-15	OLIO	Olive Oil
MOR	Morocco	COTT	Cotton Lint
TUR	Turkey	TOBA	Tobacco
MPC	Other MPCs (Algeria, Egypt, Israel, Jordan, Lebanon, Libya, Syria, Tunisia)*	WHEA	Wheat
E10	2004 EU accession countries (Cyprus, Czech Republic, Estonia, Hungary, Latvia, Lithuania, Malta, Poland, Slovakia, Slovenia)	COAR	Coarse grains (barley, maize, triticale, oats, rye, sorghum, other cereals)
BUR	Bulgaria and Romania	RICE	Rice
RUA	Russia and Ukraine	SUGA	Sugar
ANZ	Australia and New Zealand	OILS	Oilseeds
MEX	Mexico	MILK	Milk
USA	United States	BEEF	Beef and Veal
BRA	Brazil	PORK	Pig meat
CHI	China	POUL	Poultry meat
ROE	Canada, Iceland, Japan, South Korea, Switzerland, Norway		
ROW	Rest of World		

Notes: * Although Libya is not yet an official Mediterranean Partner Country of the EU, it has been included in this group of countries since it is a potential partner country and currently has observer status at the Euro-Med Agreements. Gaza Strip and West Bank have been excluded due to lack of data.

Source: Own compilation

5.1 Overview of the baseline and simulation scenarios

The simulations focus on policy shocks that take place in the EU and Mediterranean Partner Countries.

The base year of the model is 2001, when Agenda 2000 was implemented in the EU. To have a plausible representation of the existing agricultural policy regime, a Base Run scenario (BA) was necessary. To include the reforms made under Agenda 2000 for 2002 and 2003 the direct payments made for oilseeds were decreased and the direct payments made for beef were increased. In addition to the full implementation of Agenda 2000 the scenario also simulated the eastern enlargement of the EU and the implementation of the Luxembourg Agreement. The eastern enlargement is simulated by adjusting the price levels and border policies of the new member states to that of the EU-15. For the Luxembourg Agreement the option of full decoupling was chosen, since most member states chose not to apply the exemptions for coupled payments that were provided by Council Regulation (EC) 1782/2003 (Official Journal of the EU L 270). The subsequent reform of the EU sugar sector, which occurred in 2006, and the reform of the tomato market (2007) are not taken into account.

The method for implementing the decoupled direct payment scheme within the Base Run is described below. First, the total sum of all decoupled payments was calculated and this total was divided by the relevant total area, to determine the decoupled payment per hectare. The relevant total area consists of the area used for cereals (including rice), oilseeds, sugar beet, sugar cane, cotton, olive oil, tobacco and for forage. For crops, the result is equivalent to the subsidy per production activity level. For milk and beef the results were divided by the stocking density, to determine the subsidy per production activity level for the products of ruminants. Pork and poultry are assumed to not be directly affected by the decoupling of payments. The subsidy per unit of output is derived by dividing the decoupled payment per production activity level for each product by its average yield. The impact on the producer incentive price for each product is calculated by multiplying this subsidy per ton by a specific multiplier for the production-effectiveness of direct payments. It should be noted that all the subsidies in the model affect the producer incentive price (KAVALLARI et al. 2005b and c). Following CAHILL (1997) who showed that the decoupling rates for oilseeds and coarse grains without area restrictions are partial, the simulations assume a partial production-effectiveness for these products.

The reform of the CAP for Mediterranean products (cotton, olive oil and tobacco) is simulated in SC1 so as to be able to see separately the effects of decoupling the direct payments of those commodities on Mediterranean agriculture. In this scenario the direct subsidies for tobacco are fully decoupled and reduced by 50 %, for cotton and olive oil they are decoupled by 65 % and 60 % respectively.

Simulations with AGRISIM and model results 113

SC2 simulates the further enlargement of the EU with Bulgaria and Romania. The enlargement is simulated by adjusting the price level of these two countries to that of the EU. The assumptions of this scenario are taken over into the following scenarios.

In SC3 the forthcoming Free Trade Area between the EU and the MPCs is simulated. The Barcelona Agreement provides for a gradual elimination of trade barriers between the EU and the MPCs, resulting in a Free Trade Area by 2010. Here the simulation goes one step further than the agreed negotiations and assumes a customs union between the EU and the MPCs. Since the EU is a bigger player in world trade than the MPCs, the customs union is simulated by having the MPCs adapt the EU's trade policy with the rest of the world. It assumes that the MPCs apply the same trade protection as the EU, measured by the nominal protection rate. The results of this scenario can be seen as the upper bounds of the Euro-Med Agreements, while the results of SC2 represent the lower bounds[18].

SC4 and SC5 simulate two different options for liberalisation. SC4 simulates a 50 % multilateral liberalisation – a possible outcome of the ongoing WTO negotiations[19] – and SC5 simulates a 100 % multilateral liberalisation. Although the last scenario is an unlikely policy outcome, it is useful for checking the plausibility of the model results, helping us to understand the magnitude of the results obtained from other scenarios and further explain them.

Table 5.2 gives an overview of the simulated scenarios.

Table 5.2: Base run and simulated scenarios

BA	Agenda 2000, EU-east enlargement and Luxembourg Agreement
SC1	Base run + CAP Reform in Mediterranean Products (cotton, olive oil, tobacco)
SC2	SC1 + EU enlargement with Bulgaria and Romania
SC3	SC2+ extension of the Euro-Med Agreements into a customs union between the EU and the MPCs (i.e. application of the EU's Nominal Protection Rate by MOR, TUR and MPCs)
SC4	SC2+ 50 % multilateral liberalisation across the entire world
SC5	SC2 + 100 % multilateral liberalisation across the entire world

Source: Own compilation

This chapter only discusses the simulation results for the EU-27 and the MPCs. Detailed tables with the baseline projections and results on these countries can be found in the Annex (Tables B.1 to B.9). Results for the other regions of the model are not presented here but are available from the author. The results presented in section 5.1 concern changes from the Base Run Assumption (BA) on supply, demand,

---

[18] Hereafter this scenario will be called „Barcelona Agreement"

[19] Hereafter this scenario will be called „WTO liberalisation"

net trade, prices and social welfare, while in sections 5.2, 5.3 and 5.4 the changes are presented in comparison to SC2. As will be explained later, SC2 can be seen as a new Base Run scenario, since it is the closest representation of the current policy situation. This method of presentation enables direct comparison of the impact of future policy scenarios to the current situation.

5.2 Effects of the CAP Reform for Mediterranean Commodities and the Accession of Bulgaria and Romania

The reform of the CAP for Mediterranean commodities and the enlargement of the EU to include Bulgaria and Romania only slightly affect agricultural markets in the EU and have almost no impact on the markets of the MPCs.

5.2.1 Effects on EU Member States

The effects of the second wave of the CAP Reform and the enlargement with Bulgaria and Romania affect only marginally supply in the EU-25. The impacts on Bulgaria and Romania from joining the EU cannot be neglected.

Commodity balances and net trade effects

The results mostly indicate changes in the supply and demand for cotton, olive oil and tobacco, the only markets for which a policy change is simulated. Most notably the supply of cotton falls by 4 % in Spain and Greece, which are the most important cotton supplier countries in the EU. For tobacco, the decrease in supply varies between 1 % in Spain and 7 % in Italy.

Demand for olive oil decreases by about 1 % due to reform of the olive oil market, whereas the demand for cotton and tobacco remains at the same level as the Base Run scenario.

The marginal changes in the supply of cereals and livestock products due to the second wave of the CAP reform are mostly due to cross-price effects.

The CAP reforms on Mediterranean commodities and the enlargement of the EU to include Bulgaria and Romania do not change the net trade situation of the EU countries. The Mediterranean EU Member States are, generally, net exporters of these regional commodities. However, there are some exceptions Spain is a net importer of tobacco and Italy a net importer of cotton and of olive oil. Italy is by far the most important net importer of these products among EU countries, with imports coming mainly from Tunisia (COMTRADE, various years). Figure 5.1 shows the net trade effects of the CAP Reform for Mediterranean commodities on Spain, Greece and Italy. It should be noted that model calculates net trade as the difference between supply and demand. Therefore, when supply is higher than demand, a country or a

Simulations with AGRISIM and model results

region is considered to be a net exporter and when demand is higher than supply, a net importer. Table B.4 presents the effects on net trade in detail.

Figure 5.1: Net trade effects of the CAP Reform for Mediterranean commodities on Spain, Greece and Italy

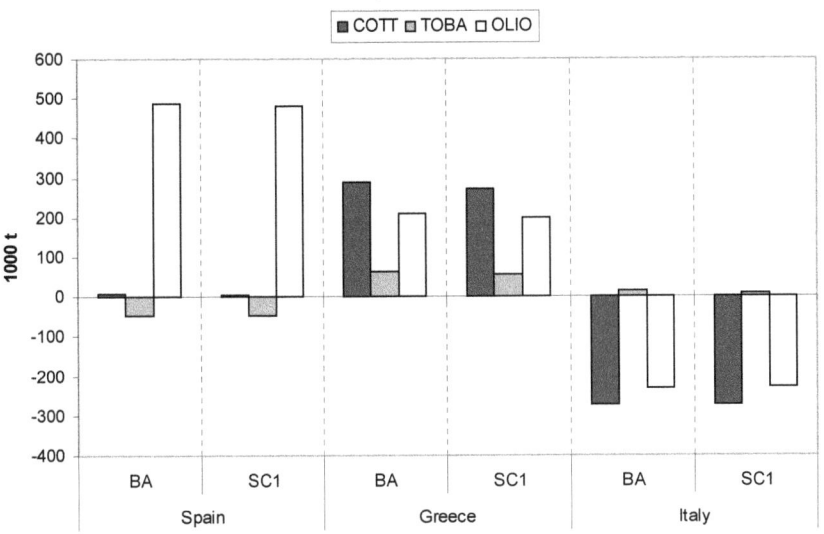

Notes: Positive values in the Y axis show net exports and negative values show net imports
Source: Own compilation based on AGRISIM simulations

The effects of the accession of Bulgaria and Romania (SC2) do not affect either supply or demand in the EU regions, and thus net trade does not change. It does though bring about changes in the commodity balances of Bulgaria and Romania.

The adjustment of price levels in Bulgaria and Romania to those of the EU-25 results in an increase in the supply of commodities that are highly protected within the EU, such as beef (increase in supply of 25 %) and cotton (an increase in supply of 33 %), all brought about by higher EU price levels and to a decrease in the supply of tomatoes by 32 %.

Changes in demand also occur in these countries, with demand for livestock commodities falling by 24 % for milk, 28 % for pork and 72 % for beef, while only for poultry meat it increases by 23 %. The changes in demand for crop products are of a lower magnitude. For example the decline in demand for cereals is between circa 1 % to 5 % and for oranges about 1 %. For olive oil, tomatoes and tobacco demand increases by 3 to 10 %.

Prices

The driving force for changes in farm gate prices for the three Mediterranean Commodities under the CAP Reform scenario is mainly due to a reduction in direct subsidies for olive oil since no change in the nominal protection rate has been simulated.

The Net Protection Rate (NPR) for the base year is taken from the PSE database of the OECD, while for the commodities and countries not included in this dataset it is calculated using the applied tariffs reported in the TRAINS database and the export subsidies reported by the WTO secretariat (Table 5.3). It should be noted that the levels of border protection are the same throughout the EU regions.

Table 5.3: Net Protection Rate in the EU-25 markets, in %

Commodity	Base Year	BA	SC1	SC2
WHEA	2	2	2	2
COAR	8	8	8	8
RICE	40	40	40	40
OILS	0	0	0	0
SUGA	75	75	75	75
MILK	60	60	60	60
BEEF	164	142	142	144
PORK	23	23	23	23
POUL	49	49	49	49
COTT	0	0	0	0
TOBA	0	0	0	0
OLIO	0	0	0	0
APPL	7	7	7	7
ORAN	15	15	15	15
TOMA	0	0	0	0

Source: Own calculations with AGRISIM

The changes in farm gate prices for tobacco are about +1 % and about +2% for olive oil, both due to adjustments in supply.

A more important influence on supply is the producer incentive price, which is estimated as the farm gate price and the part of the direct subsidies that affects production. This is the price that the farmer actually receives and the one that determines farmers' decisions about what and how much to produce. Decoupling the direct subsidies for cotton, tobacco and olive oil leads to a decline in the producer incentive prices in Spain, Greece and Italy: of 6 to 14 % for olive oil, 20 to 39 % for cotton and 19 to 27 % for tobacco. For other commodities the changes in producer incentive prices match those of farm gate prices, since the simulated scenarios do not model any further CAP reforms. Under SC1 cross-price effects lead to changes in producer incentive prices in the EU Mediterranean countries for other commodities. For example in Greece the producer incentive price for milk rises by up to 20 %.

The same trends are observed in the 2004 accession countries.

The effects of the EU enlargement to include with Bulgaria and Romania are only felt in those two countries. In both of these countries admission to the EU involves adjusting farm gate prices to the EU level and this results in a significant increase of the farm gate price of, for example, beef (by 164 %) and a decrease of the farm gate prices for cereals of 10 to 23 %.

The revenue of producers changes in response to the new price and supply levels. These changes are minor and of a small magnitude, as shown in Table B.6. Spanish and Greek cotton farmers face a revenue reduction by 4 and 3.5 % respectively. Greek tobacco producers face a loss of their revenue of 3.5%, those in Italy of 6 % and Spanish tobacco farmers one of just 0.5 %. Olive oil producers in the three different countries enjoy an increase of their revenue of between 0.1 and 0.7 %. Farmers in the rest of the EU-15 growing these crops face changes in their revenue of a similar magnitude. Farmers growing typically „northern" commodities such as cereals experience increases in their revenue of not more than 2 % (increasing in Spain and Italy, decreasing in Greece and remaining unchanged in the other EU regions). It should be noted that the magnitude of the effects would be higher for the three Mediterranean commodities if producer's incentive prices were used instead of farm gate prices to calculate farmers' revenues.

In Bulgaria and Romania farmers' revenue is adjusted as a result of their countries entrance to the EU, again depending on the changes in prices and supply. The largest change by far is an increase in beef producers' revenue, by 230 %.

The changes in border prices in the two simulations are equivalent to changes in farm gate prices as a price transmission elasticity of 1 has been assumed.

Budgetary, allocative and welfare effects

This part of the discussion focuses on the distributional and allocative effects, considering producers, quota owners, consumers, taxpayers and society at large. The effects on the taxpayers are reflected in the effects on state budgets, and the effects on other groups are shown using the surplus concept. The changes in state budgets are attributed to changes in agricultural subsidies (direct, input, general and decoupled direct subsidies) and to changes in customs duties.

The driving force for budgetary changes in the scenario reforming the CAP regime for cotton, tobacco and olive oil (SC1) are the changes in the direct subsidies made for those three commodities (Table 5.4) and particularly the decoupled direct subsidies within the EU-25 regions (Table 5.5). It should be noted that changes in the decoupled direct subsidies within the 2004 accession countries are close to zero as the

Base Run of the model assumes direct payments in these countries to have already been decoupled. Changes in the customs duties in this scenario are minimal and are only reflected in changes of the quantities traded since, as discussed above, no change in EU border protection was simulated (Table 5.6). Moreover the changes in the input and in the general subsidies are close to zero.

Table 5.4: Budgetary effects from changes in direct subsidies within the EU-25 (deviations from Base Run in US$ million)

	Cotton			Tobacco			Olive oil		
	BA	SC1	SC2	BA	SC1	SC2	BA	SC1	SC2
Greece	0	323	323	0	337	337	0	321	321
Italy	0	0	0	0	303	303	0	461	461
Spain	0	127	127	0	103	103	0	559	559
Rest of EU-15	0	0	0	0	105	105	0	21	21
2004 Accession States (EU-10)	0	0	0	0	0	0	0	0	0

Source: Own calculations with AGRISIM

Table 5.5: Budgetary effects of changes in the decoupling of direct subsidies in the EU-27 (all agricultural markets, deviations from Base Run in US$ million)

	BA	SC1	SC2
Greece	0	-799	-800
Italy	0	-608	-608
Spain	0	-730	-729
Rest of EU-15	0	-73	-73
2004 Accession States (EU-10)	0	0	0
Bulgaria & Romania	0	0	-2541

Source: Own calculations with AGRISIM

Table 5.6: Budgetary effects of changes in customs duties in the EU-25 (deviations from Base Run in US$ million)

Commodity	Greece			Italy			Spain			Rest of EU-15			EU-10		
	BA	SC1	SC2	BA	SC1	SC2	BA	SC1	SC2	BA	SC1	SC2	BA	SC1	SC2
WHEA	0	0	0	0	0	0	0	0	0	0	0	0	0	0	0
COAR	0	0	0	0	-1	0	0	-1	-1	0	0	1	0	0	0
RICE	0	0	0	0	0	0	0	0	0	0	0	0	0	0	0
OILS	0	0	0	0	0	0	0	0	0	0	0	0	0	0	0
SUGA	0	0	0	0	0	0	0	0	0	0	0	-1	0	0	0
MILK	0	0	0	0	0	0	0	0	0	0	0	8	0	0	10
BEEF	0	-1	0	0	-5	-4	0	-9	-9	0	-1	-5	0	1	10
PORK	0	0	1	0	1	2	0	2	5	0	0	14	0	0	3
POUL	0	0	0	0	0	-1	0	1	0	0	0	-7	0	0	-2
COTT	0	0	0	0	0	0	0	0	0	0	0	0	0	0	0
TOBA	0	0	0	0	0	0	0	0	0	0	0	0	0	0	0
OLIO	0	0	0	0	0	0	0	0	0	0	0	0	0	0	0
APPL	0	0	0	0	0	0	0	0	0	0	0	0	0	0	0
ORAN	0	0	0	0	0	0	0	0	0	0	0	0	0	0	0
TOMA	0	0	0	0	0	0	0	0	0	0	0	0	0	0	0

Source: Own calculations with AGRISIM

The scenario regarding Bulgaria's and Romania's entry to the EU shows changes in the budgets of the two countries, which are due to the required changes to their agricultural subsidies and their customs duties, as shown in Table 5.7. The changes caused by decoupled direct subsidies in these countries are included in Table 5.5.

In total the budgetary effects on the EU-27 are positive. They are mostly due to lower levels of expenditure on decoupled direct subsidies, which are reduced for tobacco. These benefits are partly offset by small changes in trade levies brought about by changes in the net quantity traded commodities.

Table 5.7: Budgetary effects of changes in direct subsidies and customs duties in Bulgaria and Romania (deviations from Base Run in US$ million)

Commodity	Change of direct subsidies			Change of customs duties		
	BA	SC1	SC2	BA	SC1	SC2
WHEA	0	0	0	0	0	14
COAR	0	0	0	0	0	-7
RICE	0	0	0	0	0	-52
OILS	0	0	0	0	0	-1
SUGA	0	0	0	0	0	-4
MILK	0	0	-25	0	0	-143
BEEF	0	0	0	0	0	-344
PORK	0	0	-6	0	0	-40
POUL	0	0	-2	0	0	39
COTT	0	0	0	0	0	0
TOBA	0	0	0	0	0	-7
OLIO	0	0	0	0	0	-1
APPL	0	0	0	0	0	2
ORAN	0	0	0	0	0	1
TOMA	0	0	0	0	0	-32

Source: Own calculations with AGRISIM

To capture the budgetary effects on the different EU regions a new module was programmed for the model. Initially the taxpayers' or budgetary effects were calculated by adding the changes in trade levies and agricultural subsidies (differentiated into input, direct, general and decoupled subsidies) and assuming these attributes as additional expenditure or revenue (depending on the deviation from the Base Run) for the national budget. However, following PUSTOVIT (2003, p. 159), the intra-community financial flows between the Member States' and Brussels also need to be taken into account. Thus, the agricultural subsidies included in the model are financed by the EU, and the customs duties are collected by each member state on behalf of the EU and flow to the EU budget. Additionally, each member state makes a contribution to the EU budget which is based on its national contribution i.e. the GNI-based (Gross National Input) its own resources payments, the VAT-based (Value Added Tax) own resources payments and the UK rebate. Table 5.8 shows the changes to the budget of the EU-25 regions resulting from the CAP Reform for Mediterranean commodities and the EU's enlargement with Bulgaria and Romania. The intra-community financial flows benefit southern EU Member States which make a smaller contribution to the EU budget. For these countries the budgetary effects are brought about by changes in revenues, while for the rest of EU-15 the effects are due to expenditure changes, since they contribute more to the EU's budget. For the ten new Member States the changes are also related to expenditure since they collect now customs duties that go to the EU budget.

Table 5.8: Budgetary changes in the EU-25 due to the CAP Reform for Mediterranean commodities (SC1) and the accession of Bulgaria and Romania (SC2) (in US$ million)

Region	Financing System	BA	SC1	SC2	Deviations from BA	
					SC1	SC2
Spain	National	-6669	-6618	-6615	51	54
	Intra-community	5059	5047	5047	-12	-12
Greece	National	-5848	-5666	-5665	182	183
	Intra-community	5513	5340	5340	-174	-174
Italy	National	-5952	-5801	-5799	151	153
	Intra-community	3304	3217	3220	-87	-84
Rest of EU-15	National	-6972	-6920	-6907	52	65
	Intra-community	-6080	-5821	-5805	259	275
EU-10	National	7188	7188	7209	0	21
	Intra-community	-7797	-7783	-7802	14	-5

Source: Own calculations based on AGRISIM simulations and on EUROPEAN COMMUNITIES, 2005

The two simulated scenarios result in a reduction of the budget revenue of the EU's Mediterranean member states when intra-community financial flows are taken into account because of the custom duties effects. Thus the deviations of SC1 and SC2 from the Base Run are negative, with a reduction of US$12 million in Spain, US$174 million in Greece and US$84 and 87 million in the two scenarios in Italy. However, if these changes in financial flows would place a burden on these countries' national budgets, then their budgetary expenditure would be reduced, as shown in the columns with the deviations from BA and the effect on taxpayers, expressed as deviations to the Base Run, would be positive. For the rest of the EU-15 the deviations from the Base Run due to the CAP Reform of Mediterranean commodities and the EU enlargement with Bulgaria and Romania are positive and are of higher magnitude when intra-community financial flows are taken into consideration. In the ten new Member States the financial flows within the EU imply an increase in budgetary expenditure. This is despite their low contribution to the Brussels budget and is due to the customs duties that they collect on behalf of the EU being higher than the agricultural subsidies paid for out of Brussels' budget. SC1 results in lower expenditure for these countries, when adjusted for intra-community financial flows (lower expenditure) but slightly higher expenditure under SC2, which takes into account the EU's enlargement with Bulgaria and Romania.

Figures 5.2 and 5.3 illustrate the welfare effects on producers, quota owners and consumers among the EU-25 regions.

Figure 5.2: Allocation of welfare effects among the EU Mediterranean Member States due to the CAP Reform for Mediterranean commodities (SC1) and the accession of Bulgaria and Romania (SC2) (deviations from BA)

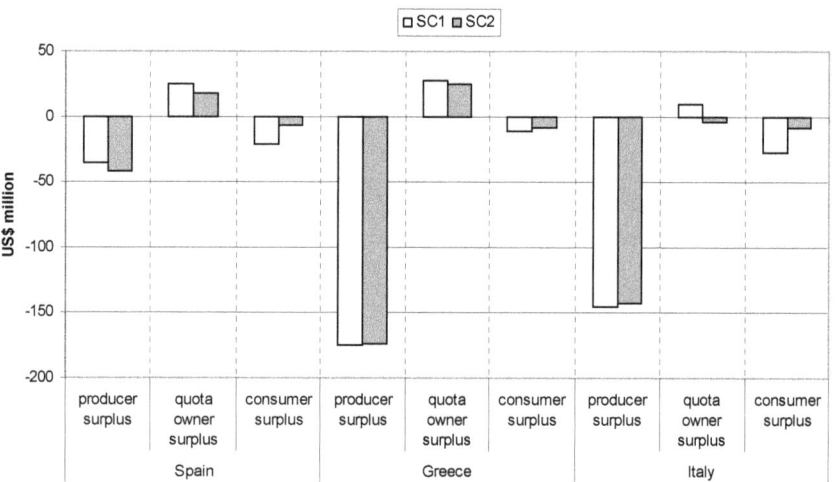

Source: Own compilation based on AGRISIM simulations

Figure 5.3: Allocation of welfare effects on the rest of the EU-15 and 2004 accession states as a result of the CAP Reform of Mediterranean commodities (SC1) and the accession of Bulgaria and Romania (SC2) (deviations from BA)

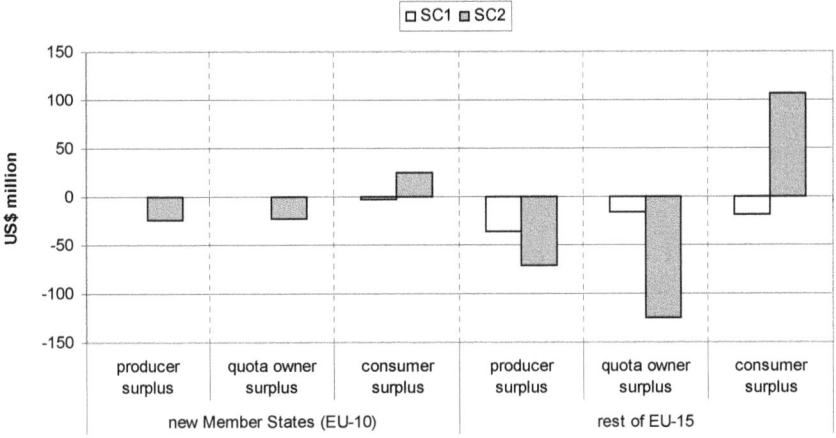

Source: Own compilation based on AGRISIM simulations

Producer surplus is reduced in all EU-25 regions, with the largest reductions being in Greece and Italy (US$175 and US$145 million respectively under both SC1 and SC2). This is partly because the reductions in the farm gate prices for cotton, tobacco and olive oil in these countries are greater than those in the other EU regions and partly because these commodities have a larger share in the commodity composition of these two countries than in the other EU regions.

The effects on quota owners and on consumers are minor. For quota owners the effects are attributed to cross price effects for commodities to which production quotas are applied, such as sugar and milk. For consumers they are due to the small effects that these changes have on demand.

The overall welfare effects on the EU regions depend upon the budgetary effects. Due to intra-community financial flows, the changes in welfare are negative for Greece, Italy and Spain, where the revenue decreases, but are positive for the rest of the EU-15 and the ten new Member States. In the EU-25 as a whole the CAP reform of Mediterranean commodities and the enlargement with Bulgaria and Romania result in increases in welfare of US$ 10 and 22 million respectively. Table B.9 in the Annex presents the results in detail.

These results show that the effects of the EU enlargement with Bulgaria and Romania on the EU-25 are only marginal. However, the welfare effects of acceding to the EU are of a considerable magnitude for Bulgaria and Romania, although those of SC1 are negligible (Figure 5.4). Producers in these two countries will enjoy welfare gains of about US$2,527 million as a result of the adjustment of price levels to those of the EU, while there is a slight decrease in consumer welfare as a result of an adjustment of demand. For Bulgaria and Romania the intra-community financial flows system of the EU was modelled, since their exact contribution to the EU budget was not known at the time that the module was programmed. Instead budgetary effects are illustrated under the assumption that these countries would have to finance the application of EU agricultural policy from their own resources. The overall negative welfare effects are a result of this assumption.

Figure 5.4: Welfare effects on Bulgaria and Romania due to the CAP Reform for the Mediterranean commodities (SC1) and their accession to the EU (SC2) (deviations from BA)

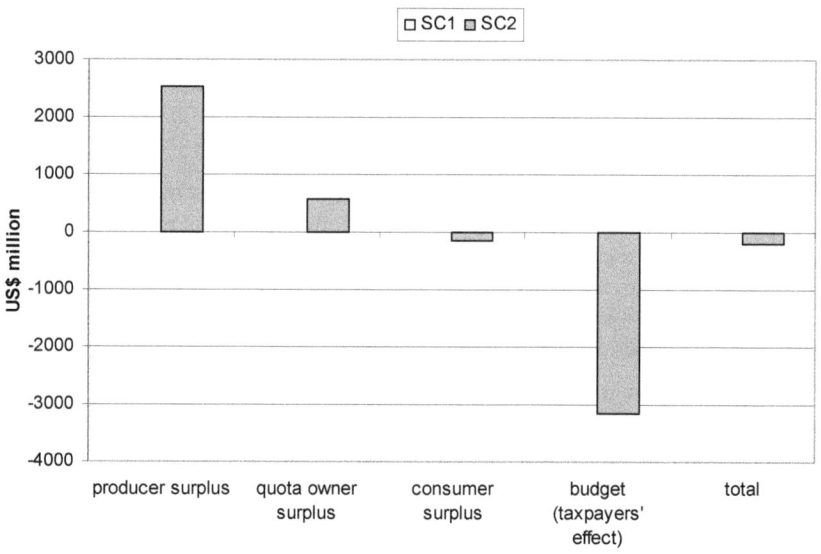

Source: Own compilation based on AGRISIM simulations

5.2.2 Effects on the MPCs

The effects on the Mediterranean Partner Countries of the CAP reforms for cotton, olive oil and tobacco and the EU enlargement by Bulgaria and Romania are minimal, with the changes in the commodity balances, prices and social welfare being almost close to zero. However, the simulations for those two scenarios referred to shocks only on the EU's markets and any effects on the MPCs would result from the market clearing mechanism of the model and would be due to chain effects on the rest of the world, stemming from changes in the EU's markets. Despite the EU's dominant position as global player and as the most important player in the Mediterranean region, the results indicate that the MPCs were not affected by the change in market support for cotton, olive oil and tobacco or by the enlargement of the EU to include Bulgaria and Romania. This could be also explained by the small effects that SC1 and SC2 have on the EU's markets, as discussed above.

Commodity balances and net trade effects

The only distinguishable effect of the reform of the CAP regime for cotton, olive oil and tobacco and the enlargement of the EU with Bulgaria and Romania is an increase in the supply of olive oil by 1 % in Turkey and in the rest of the MPCs.

Demand remains at the same levels as in the Base Run.

As a result there is almost no change in the net trade and the trade status of the MPCs. They remain net importers of typically "northern" commodities, such as cereals, sugar, oilseeds and livestock products (meat and dairy products) and net exporters of olive oil, tomatoes and oranges. In detail, Morocco is a net exporter of about 260,000 t of oranges in both SC1 and SC2 and of about 207,000 t and 210,000 t of tomatoes in SC1 and SC2 respectively, while its net exports of olive oil are only about 2,000 t. Turkey is a net exporter of about 108,000 t of olive oil and 143,000 t of oranges in both SC1 and SC2. In SC2, Turkey's net exports of tomatoes increase by 45,000 t (from 642,000 t to 687,000 t). Turkey is also a net exporter of sugar, with net exports of about 900,000 t. The rest of the MPCs are net exporters of olive oil (about 107,000 t) and oranges (about 430,000 t) in both scenarios. Net exports of tomatoes from these countries also increase under SC2, by 38,000 t (from about 99,000 to almost 137,000 t). The rest of MPCs are also net exporters of cotton (about 183,000 t), largely from Egypt.

Prices

The price effects are minimal and are due to the changes in supply, since no changes in the price level or border protection of the MPCs was simulated. Table 5.9 presents the applicable net protection rates for the two first scenarios.

In detail, the CAP Reform for Mediterranean commodities results in a marginal rise in farm gate prices which is highest for olive oil (about 2 %) and tobacco (about 1 %) in all three MPC regions in the model, i.e. in Morocco, Turkey and the rest of the MPCs.

The changes in farm gate prices are transmitted to the border prices with a price transmission elasticity of 1. Since domestic policy within the three MPCs is not modelled the producer incentive price is the same as the farm gate price.

Table 5.9: Net protection rates in the MPCs, in %

Commodity	Morocco				Turkey				Rest of MPCs			
	Base Year	BA	SC1	SC2	Base Year	BA	SC1	SC2	Base Year	BA	SC1	SC2
WHEA	29	29	29	29	-7	-7	-7	-7	7	7	7	7
COAR	0	0	0	0	19	19	19	19	0	0	0	0
RICE	111	111	111	111	0	0	0	0	0	0	0	0
OILS	0	0	0	0	25	25	25	25	0	0	0	0
SUGA	0	0	0	0	27	27	27	27	2	2	2	2
MILK	115	115	115	115	16	16	16	16	26	26	26	26
BEEF	0	0	0	0	207	207	207	207	8	8	8	8
PORK	0	0	0	0	0	0	0	0	2	2	2	2
POUL	109	109	109	109	19	19	19	19	20	20	20	20
COTT	3	3	3	3	0	0	0	0	0	0	0	0
TOBA	18	17	18	18	25	25	25	25	4	4	4	4
OLIO	55	54	55	55	31	31	31	31	43	43	43	43
APPL	55	54	55	55	0	0	0	0	7	7	7	7
ORAN	0	0	0	0	0	0	0	0	7	7	7	7
TOMA	0	0	0	0	0	0	0	0	37	37	37	37

Source: Own calculations with AGRISIM

Budgetary, allocative and welfare effects

The budgetary effects are also very small, coming about solely as a result of changes in customs duties, as shown in detail in Table 5.10. These changes, as explained above, are due to small changes in the supply, which in turn are a chain effect of the changes in the EU's markets.

The effects show that the resources are equally allocated between producers and consumers throughout the MPCs (Table B.9 in the Annex). The slight increase in farm gate prices results in an increase in the producer surplus, which is in Morocco of about US$3 and US$2 million in both SC1 and SC2. Turkey experiences an increase in producer surplus of about US$8 under SC1, but no observable change under SC2. The rest of the MPCs see an increase of about US$20 and US$17 million under SC1 and SC2 respectively. This change in price levels leads to a decrease of the consumer surplus, which is almost of the same magnitude as the increase in the producer surplus.

Overall welfare is only slightly affected. In Turkey, there are welfare gains of about US$2 million from the EU enlargement to include Bulgaria and Romania and a similar level of welfare losses under the CAP reform for Mediterranean commodities, both attributed to the changes of the customs duties. In the rest of the MPCs the welfare losses due to SC1 and SC2 are about US$6 and US$4 million respectively, while in Morocco the overall welfare effects are close to zero.

Table 5.10: Changes in the customs duties of the MPCs due to the CAP reform of Mediterranean commodities (SC1) and the accession of Bulgaria and Romania to the EU (SC2) (deviations from Base Run in US$ million)

Commodity	Morocco			Turkey			Rest of MPCs		
	BA	SC1	SC2	BA	SC1	SC2	BA	SC1	SC2
WHEA	0	0	0	0	0	0	0	0	0
COAR	0	0	0	0	0	0	0	0	0
RICE	0	0	0	0	0	0	0	0	0
OILS	0	0	0	0	0	0	0	0	0
SUGA	0	0	0	0	0	0	0	0	0
MILK	0	0	0	0	0	1	0	0	1
BEEF	0	0	0	0	0	2	0	0	0
PORK	0	0	0	0	0	0	0	0	0
POUL	0	0	0	0	0	0	0	0	-1
COTT	0	0	0	0	0	0	0	0	0
TOBA	0	0	0	0	0	0	0	0	0
OLIO	0	0	0	0	-1	-1	0	-4	-4
APPL	0	0	0	0	0	0	0	0	0
ORAN	0	0	0	0	0	0	0	0	0
TOMA	0	0	0	0	0	0	0	0	-5

Source: Own calculations with AGRISIM

Thus the effects of SC1 and SC2 on the MPCs are only marginal. Since SC2 refers to the current policy regime, it can be taken as the new Base Run scenario. In the following sections the effects will expressed as deviations from SC2 and are thus expressed in percentage points. References to the scenarios are kept as in Table 5.2, so as to avoid confusion.

5.3 The effects of the „Barcelona Agreement"

This section focuses on the effects of SC3 and the deviations are given in comparison to the existing policy regime (SC2) for the reasons given above.

This scenario aims to model the effects of the Euro Med Agreements. Because the model treats the commodities as homogenous and does differentiate them on the basis of their origin, the simulation models a customs union between the EU and the MPCs rather than a Free Trade Area between the two groups of countries. Moreover, because the EU is the dominant player in agricultural trade, it is assumed that the MPCs will apply the same border protection as the EU. Thus the simulation is accomplished by applying the EU's NPR to the MPCs.

5.3.1 Effects on the EU Member States

Commodity balances and net trade effects

The results indicate that the „Barcelona Agreement" does not result in any distinguishable changes in the commodity balances in the EU regions. Olive oil production in Spain and Greece increases by one percentage point (deviation of SC3 from SC2), the supply of tomatoes increases by three percentage points in each of Spain, Greece and the 2004 accession states, by one percentage point in the rest of the EU-15 and by two percentage points in Italy. The supply of oranges from Italy, the 2004 accession states and Greece declines, by almost one percentage point. For all other commodities, the changes bought about by the „Barcelona Agreement" deviate by less than one percentage point from the new Base Run (SC2).

Demand remains at the same levels as under SC2, with no observable changes from to the application of the EU's NPR by the MPCs.

These changes in demand and supply are reflected in the changes in net trade. Figure 5.5 shows the changes in net trade in detail. Due to the almost marginal changes in supply, the changes in the quantities traded are very small and there is no change in net trade status. The sole exception to this is the net trade in poultry meat in Spain, where the simulation of the „Barcelona Agreement" changes Spain's trade status from being a net exporter of 1,000 t of poultry meat to a net importer, of 5,000 t.

The net trade in specific commodities, such as tomatoes and olive oil develops in favour of the Mediterranean EU Member States, which export a little more and to import a little less than under the current agricultural policy regime (new Base Run or SC2). For example the „Barcelona Agreement" leads to an increase in net exports of tomatoes from Spain and Italy by 100,000 t each and an increase of 50,000 t of net exports of tomatoes from Greece. Net exports of olive oil from Spain and Greece increase by 20,000 t, and Italy's net imports of olive oil decrease by approximately the same magnitude. The „Barcelona Agreement" has negative effects on exports of oranges from these three countries, with net exports decreasing by 10,000 t in each country. The other commodity market where distinguishable changes occur is wheat, with Spain and Italy decreasing their net imports by 20,000 t each, and Greece remaining a net exporter of 95,000 t. Finally the Agreement does not affect the net trade of cotton, commodity important only for Greece and Spain.

Figure 5.5: Net trade effects of the „Barcelona Agreement" on Spain, Greece and Italy (SC3)

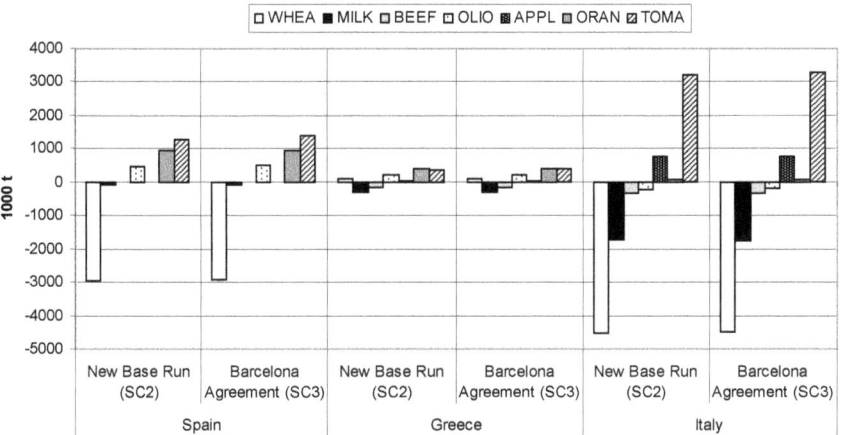

Notes: Positive values in the Y axis show net exports, negative values show net imports
Source: Own compilation based on AGRISIM simulations

For the rest of EU-15, the „Barcelona Agreement" leads to an increase in net exports of wheat that are in the order of 180,000 t, a decrease of exports of coarse grains by 120,000 t and a decrease of tomato imports, by 40,000 t. This decline in the net imports of tomatoes suggests that the additional tomato exports from Greece, Italy and Spain (discussed above) would not be distributed to EU markets but would be exported to the rest of the world (Figure 5.6).

For the ten 2004 accession countries, and Bulgaria and Romania the changes brought about by the „Barcelona Agreement" are very minor. The most distinguishable changes are a decline of 40,000 t in the net exports of coarse grains from the ten 2004 accession countries, and a decline in the net exports of milk of Bulgaria and Romania, which drop by 70,000 t (Figure 5.6).

Figure 5.6: Net trade effects of the „Barcelona Agreement" on non-Mediterranean EU regions (SC3)

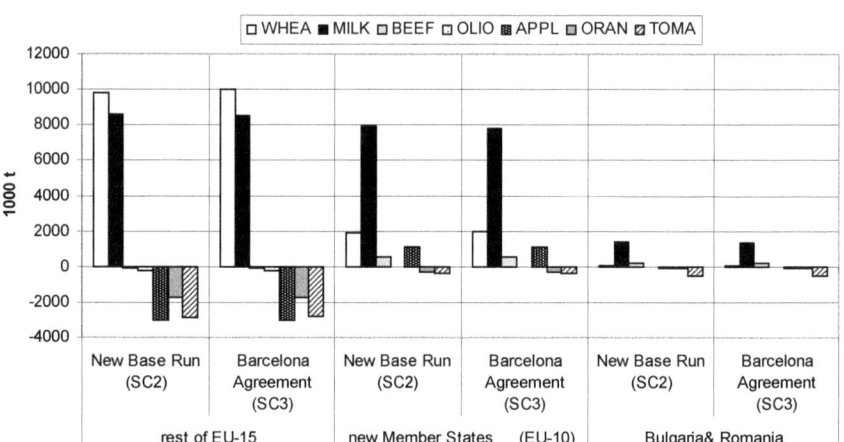

Notes: Positive values in the Y axis show net exports, negative values show net imports
Source: Own compilation based on AGRISIM simulations

Prices

The price effects brought about by the „Barcelona Agreement" occur solely as a result of changes in the commodity balances within EU markets. To remind the reader, in this scenario the only adjustment made has been to change the NPR of the MPCs to the EU level. No further shocks or changes to the EU's markets were modelled. Border protection, expressed as the difference between border prices and domestic prices has been kept at the level of SC2, which is modelled through the NPR, and there are no additional changes in the implementation of the CAP. An overview of the applied NPR in the EU's markets up to SC3 is presented in Table 5.11.

This is why the deviations in farm gate prices between SC3 and the new Base Run (SC2) are within +/-1% for all commodity markets in the model, with the exception of olive oil, where the effects are slighty higher. Here SC3 leads to an increase of 4, 6, 5 percentage points in farm gate prices in Spain, Italy and Greece respectively and 3 percentage points in each of the rest of the EU regions (EU-15, 2004 accession countries and Bulgaria and Romania), reflecting the inelastic supply of olive oil (the own-supply elasticity of olive oil in Spain, Greece and Italy is of 0.15, 0.17 and 0.08 respectively, while in the rest of EU-15 and in EU-10 of 0.16 and 0.15 respectively).

Simulations with AGRISIM and model results

The changes in border prices are of the same magnitude as the changes in farm gate prices throughout the markets and the EU regions due to the assumed price transmission elasticity, as explained above.

Table 5.11: Net protection rates in the EU-25 markets, in %

Commodity	Base Year	BA	SC1	SC2	SC3
WHEA	2	2	2	2	2
COAR	8	8	8	8	8
RICE	40	40	40	40	40
OILS	0	0	0	0	0
SUGA	75	75	75	75	75
MILK	60	60	60	60	60
BEEF	164	142	142	144	150
PORK	23	23	23	23	23
POUL	49	49	49	49	49
COTT	0	0	0	0	0
TOBA	0	0	0	0	0
OLIO	0	0	0	0	0
APPL	7	7	7	7	7
ORAN	15	15	15	15	15
TOMA	0	0	0	0	0

Source: Own calculations with AGRISIM

The „Barcelona Agreement" only has minor effects on farmers' revenue, and these are only distinguishable within the olive oil and tomato markets. Olive oil producers enjoy an increase in their revenue of five percentage points, with Greek olive oil producers being best of all with an increase in their incomes of seven percentage points. In the new Member States (EU-10 and Bulgaria & Romania) the increase in revenue is three percentage points. Tomato farmers also benefit from an increase in their revenue, varying between four percentage points in Spain, Greece, the rest of EU-15 and the ten new Member States to 2.5 percentage points in Italy and 1.5 percentage points in Bulgaria and Romania. In Bulgaria and Romania there are some disparities, with the revenue of sugar and beef producers, decreasing by 3.5 and 5.0 percentage points respectively.

Budgetary, allocative and welfare effects

In the simulation of the „Barcelona Agreement" the driving force for the budgetary effects are the changes in customs duties which are a direct result of changes in the quantities traded. The border protection and expenditure on agricultural subsidies in the EU are kept at the same level as in SC2. The effects are of a small magnitude, which is to be expected because of the marginal changes in net trade. The revenue changes in the rest of the EU-15 stem from changes in the milk and poultry markets and in the ten new Member States they are due to changes in the milk and beef markets (Table 5.12).

Table 5.12: Change in customs revenues in the EU-27 due to the „Barcelona Agreement" (SC3) (deviations from new Base Run (SC2) in US$ million)

Commodities	Greece	Italy	Spain	rest of EU-15	new Member States (EU-10)	Bulgaria & Romania
WHEA	0	0	0	-1	0	0
COAR	0	0	0	-1	0	0
RICE	0	0	0	0	0	0
OILS	0	0	0	0	0	0
SUGA	0	0	0	5	1	0
MILK	0	0	1	24	28	8
BEEF	2	5	0	1	22	7
PORK	0	0	1	4	0	0
POUL	0	2	2	11	1	0
COTT	0	0	0	0	0	0
TOBA	0	0	0	0	0	0
OLIO	0	0	0	0	0	0
APPL	0	0	0	0	0	0
ORAN	1	1	1	0	0	0
TOMA	0	0	0	0	0	0

Source: Own calculations with AGRISIM

Table 5.13 shows the welfare effects and the allocation of resources between producers, consumers and taxpayers, all expressed as deviations from SC2. Producers in the EU-15 region enjoy welfare gains from the implementation of the „Barcelona Agreement" because of a slight increase in farm gate prices, while in the ten new Member States the producer surplus decreases, due to a slight reduction of the farm gate prices for livestock.

The slight changes in farm gate prices are the driving force for changes in consumer surplus. Overall there is a decrease in consumer surplus in the three EU Mediterranean member states and an increase in the other EU regions.

The quota owner surplus effects are attributed to the changes in the milk, sugar and tobacco markets and are more evident in the rest of the EU-15 and in the twelve EU Member States (EU-10, Bulgaria and Romania) because these countries have the highest share of milk and sugar quotas.

Table 5.13: Welfare effects in the EU-27 due to the „Barcelona Agreement" (SC3) (deviations from new Base Run (SC2) in US$ million)

	Spain	Greece	Italy	rest of EU-15	new Member States (EU-10)	Bulgaria & Romania
producer surplus	135	55	99	39	-24	0
quota owner surplus	-24	-7	-42	-355	-70	-20
consumer surplus	-32	-27	-44	249	45	6
budget[1]	3	2	8	43	53	15
total[1]	81	23	21	-25	4	1
budget[2]	6	0	8	35	-49	–
total[2]	85	21	22	-33	-98	–

Notes: [1] Without considering intra-community financial flows; [2] Considering intra-community financial flows
Source: Own calculations with AGRISIM

The regional budgetary effects are small and positive for the EU-15 region, whether or not intra-community financial flows are taken into account. For the ten 2004 accession states the intra-community financial flows have negative budgetary effects (as was the case for SC1 and SC2) since import tariffs are collected on behalf of the EU and do not flow into the national budget. The intra-community financial flows between the EU and Bulgaria and Romania were not modelled as they were not known at the time.

Overall the scenario of the „Barcelona Agreement" results in welfare gains in the southern EU member states and welfare losses in the northern ones (the rest of the EU-15 and the EU-10). The overall welfare gains for the EU-25 are US$105 million.

5.3.2 Effects on the MPCs

The effects of the „Barcelona Agreement" on the MPCs are more pronounced, although in general they are also of small magnitude. This is to be expected because the simulations only involve shocks on the border protection of the markets of these countries. The following section contains detailed results of the effects on commodity balances, prices, state budgets, allocation and welfare.

Commodity balances and net trade

The effects on commodity balances are in line with the changes in the NPR levied by the MPCs and depend on the initial net trade status of each country. Due to the different initial protection rates in the MPC regions the effects will differ from country to country.

In Morocco the wheat supply decreases by 6 percentage points, the supply of other cereals such as coarse grains and of rice decreases by eleven and six percentage points respectively. This is because the EU border protection level is lower than that currently applied by Morocco. For the same reason oilseeds production decreases by one percentage point, the milk supply by six percentage points, the poultry supply by 23 percentage points and the supply of apples by 26 percentage points. The beef supply increases by 23 percentage points and the supply of sugar increases by 6 percentage points, both due to the application of a higher NPR. More significantly there are changes in the supply of typical Mediterranean commodities, with an increase in the supply of tomatoes, oranges and cotton, by 4, 11 and 7 percentage points respectively, and a decrease of the supply of tobacco and olive oil by 2 and 11 percentage points respectively.

In Turkey the largest changes are observed in the markets for coarse grains, where the supply drops by 12 percentage points, and in the markets for milk, poultry meat and oranges, where production increases by 25, 12 and 11 percentage points, respectively. There is also a notable change in the tomato supply, which increases by 6 percentage points. In the other commodity markets the deviations are within + /– 5 percentage points.

In the rest of the MPCs the highest deviations occur in the tomato market, where the supply decreases by 40 percentage points, in the milk and beef market, where production increases by 12 and 20 percentage points respectively and in the olive oil markets, where there is a decrease in production by 11 percentage points.

The changes in total demand are smaller than those for the supply and are only observable in some particular markets. In Morocco and in the rest of the MPCs the demand for beef declines by 24 and 22 percentage points respectively. By contrast the demand for apples in Morocco increases by 18 percentage points and the demand for tomatoes in the rest of the MPCs increases by 13 percentage points.

Changes in net trade result from the changes in supply and demand. Figure 5.7 illustrates the net trade effects on particular commodities in the MPCs as a result of the application of the „Barcelona Agreement". The net trade status of MPCs in tomatoes changes, with these countries moving from being net exporters of 137,000 t in SC2 to net importers of 4.5 million t. Turkey and the MPCs both shift from being net importers of milk to net exporters. Turkey was previously a net importer of 181,000 t but becomes a net exporter of 2.5 million t and the rest of the MPCs change from being a net importer of 1.4 million t to a net exporter of 475,000 t. Morocco experiences changes in its net trade status for olive oil and apples, but these changes are of a smaller magnitude. The „Barcelona Agreement" leads Morocco to become a net importer of 8,000 t of olive oil (from a net exporter of 2,000 t in SC2) and a net importer of 86,000 t of apples (from a net exporter of 8,000 t in SC2). A further shift in Morocco is in the market for beef, where it shifts from a net importer to a net exporter of 70,000 t.

The effects on the rest of the markets are only marginal and the MPCs maintain their initial net trade status.

Figure 5.7: Net trade effects of the „Barcelona Agreement" on MPCs (SC3)

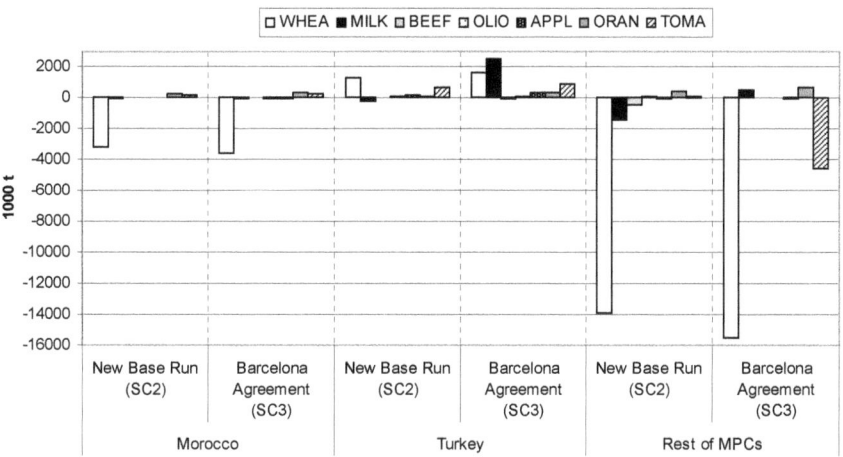

Notes: Positive values in the Y axis refer to net exports, negative values to net imports
Source: Own compilation based on AGRISIM simulations

Prices

The driving force for the price effects under SC3 is the adjustment of the protection levied by the MPCs' to the EU level. As Table 5.14 shows, there are significant disparities both among the MPCs and among the various agricultural markets in the initial level of the NPR and therefore in the effects that this change has. For example in Morocco the NPR for milk is rather high, 115 %, while in Turkey and in the rest of MPCs it is much lower, at about 20 %. This explains why the application of the EU's protection rate leads to different result for the same markets among the MPCs and why it is difficult to draw out general trends for the MPCs as a whole.

To remind the reader, the NPRs for the Base Year are derived from the PSE database of the OECD. For commodities and countries not included in this dataset the NPR is calculated by comparing the applied import tariff with the export subsidies. The applied import tariffs are expressed as *ad-valorem* equivalents and the export subsidies are derived from the TRAINS database and the WTO country notifications respectively. Thus, the entry price system of the EU for specific commodities, such as tomatoes, has not been explicitly modelled but has been indirectly taken into account through the applied AVE of import tariffs reported in the TRAINS database.

Table 5.14: Net protection rates in MPCs, in %

Commodity	Morocco		Turkey		Rest of MPCs	
	Base Year	SC3	Base Year	SC3	Base Year	SC3
WHEA	29	2	-7	2	7	2
COAR	0	8	19	8	0	8
RICE	111	40	0	40	0	40
OILS	0	0	25	0	0	0
SUGA	0	75	27	75	2	75
MILK	115	60	16	60	26	60
BEEF	0	164	207	164	8	164
PORK	0	23	0	23	2	23
POUL	109	49	19	49	20	49
COTT	3	0	0	0	0	0
TOBA	18	0	25	0	4	0
OLIO	55	0	31	0	43	0
APPL	55	7	0	7	7	7
ORAN	0	15	0	15	7	15
TOMA	0	0	0	0	37	0

Source: Own calculations with AGRISIM

The application of the EU's protection level mainly influences farm gate prices, which adjust to the EU level increasing or decreasing according to the difference between national levies and the newly imposed NPR. The border prices remain almost unaffected since no significant change in the world market prices occurs due to the simulations.

The highest disparities in farm gate prices are observed in Morocco since its border protection show the highest deviations from the EU one. Compared to SC2 (new Base Run), the farm gate prices for beef increase by 160 percentage points and those of sugar by 75 percentage points The farm gate prices for oranges and tomatoes rise by 14 and 2 percentage points respectively. By contrast, the farm gate prices for olive oil and apples decrease by 30 percentage points each. Those for rice and wheat decrease by 20 and 35 percentage points respectively.

In Turkey the deviations in farm gate prices are generally within 40 percentage points. The beef farm gate prices decrease by about 16 percentage points since Turkey applies a higher protection for beef than the EU. A decrease of 20 percentage points is observed in the tobacco, olive oil and oilseeds markets, while there is an increase of 40 percentage points in the farm gate prices for sugar and milk and one

Simulations with AGRISIM and model results

of 25 percentage points for each of pork and poultry. The farm gate prices of oranges and tomatoes increase by 14 and 4 percentage points respectively.

In the rest of the MPCs the highest increases are observed in the farm gate prices of beef sugar, rice, milk, poultry and pork of 142, 70, 40, 26, 24 and 20 percentage points respectively. For typical Mediterranean commodities, the „Barcelona Agreement" scenario results in a decrease in the farm gate prices of olive oil and tomatoes by 28 and 27 percentage points, respectively, while farm gate price of oranges is increased by 7 percentage points.

The changes in producer incentive prices are the same as for farm gate prices, since no direct subsidies are modelled for these countries. Farmers' revenue therefore adjust in response to the new farm gate prices and supply level. The decline in farm gate prices leads to a decrease in farmer's revenue for most commodities in the model (which are approximated by the changes in farmers' revenue), as illustrated in detail in Table B.6.

Moroccan farmers are particularly affected by the „Barcelona Agreement" with observable reductions in the revenue of wheat, rice, milk and poultry producers by 25, 38, 30 and 45 percentage points respectively. Reductions in farmers' revenue of 40 and 55 percentage points are observed for olive oil and apple producers. Beef producers benefit the most, with their revenue rising by about 220 percentage points. Sugar and orange producers enjoy revenue increases of 85 and 26 percentage points. The effects on tomato producers are of a lower magnitude, with their revenue growth not exceeding 7 percentage points.

In Turkey the revenue of wheat, rice, sugar, milk, pork and poultry farmers is expected to rise by 12, 47, 41, 70, 23 and 39 percentage points respectively while producers of apples, tomatoes and oranges will enjoy an increase in their revenue of 10 and 25 percentage points respectively. Farmers producing coarse grains, oilseeds, beef, tobacco and olive oil all suffer from revenue decreases of 20 percentage points.

In the other MPCs producers of wheat, tobacco, olive oil, apples, oranges and tomatoes will suffer revenue losses of 6, 3, 33, 10.5 (for both apples and orange producers) and 55 percentage points respectively. Other farmers will gain significantly more, with the revenue of rice, sugar, milk, beef, pork and poultry producers increasing by 50, 80, 40, 190, 25 and 35 percentage points respectively.

Budgetary, allocative and welfare effects

The changes in the levels of border protection and subsequent changes in net trade and domestic price levels in the MPCs will have knock-on effects on the customs duties collected. These changes will impact on the states' budgets, as shown in detail in Table 5.15. The largest changes occur in the markets where the net trade effects are the highest or where the net trade status quo changes, as for example in the beef market in the other MPCs and in Morocco and the milk market in Turkey and the other MPCs.

Table 5.15: Changes of customs duties in MPCs due to the „Barcelona Agreement" (SC3) (deviations from new Base Run (SC2) in US$ million)

Commodity	Morocco	Turkey	Rest of MPCs
WHEA	-137	-17	-93
COAR	17	18	86
RICE	0	12	-33
OILS	0	-72	0
SUGA	102	-119	324
MILK	16	-271	-358
BEEF	-224	34	-241
PORK	0	0	1
POUL	17	-19	-87
COTT	-1	0	0
TOBA	-4	19	-15
OLIO	2	48	85
APPL	4	-9	0
ORAN	-17	-15	-20
TOMA	0	0	18

Source: Own calculations with AGRISIM

Overall the adjustments to the protection levied by the MPCs to bring it in line with that of the EU leads to a reduction in trade levies and, therefore to negative budget changes. In Morocco, the burden for the taxpayers is in the order of US$225 million, in Turkey US$392 million and in the rest of the MPCs US$333 million (Table 5.16).

In all the MPCs the allocation of resources is in favour of producers, who will benefit from the implementation of the Barcelona Agreement. This is due to an overall rise in farm gate prices. This goes along with a negative effect on the demand and leads to a loss of consumer surplus. In Turkey and in the rest of the MPCs consumer surplus is reduced by roughly US$1.0 billion and US$5.5 billion respectively (Table 5.16). In Morocco the effect on consumer surplus will be marginally positive.

In total the welfare will decrease by US$75 million in Morocco, US$62 million in Turkey and by about US$0.8 billion in the rest of the MPCs.

Table 5.16: **Welfare effects in MPCs due to the „Barcelona Agreement" (SC3) (deviations from new Base Run (SC2) in US$ million)**

	Morocco	Turkey	Rest of MPCs
producer surplus	122	1332	4995
quota owner surplus	0	0	0
consumer surplus	28	-1002	-5477
budget	-225	-392	-333
total	-75	-62	-814

Source: Own calculations with AGRISIM

5.4 The effects of multilateral liberalisation

This section discusses the effects of a possible WTO liberalisation (SC4) and the effects of a full multilateral trade liberalisation (SC5) comparing them to the new Base Run. The effects of SC5 are in most cases linear i.e. of twice the magnitude of SC4, with the exception of markets which had almost no initial protection. Although this scenario is a hypothetical one, the simulations and its results are included in the report to facilitate checking the plausibility of the model and to exhibit the relative magnitude of the model's results. It can also serve as a basis for comparing this modelling exercise with others that have preceded or will follow it.

5.4.1 Effects on the EU Member States

The changes bought about by the „WTO liberalisation" scenario or even full trade liberalisation are not of a high magnitude but are still higher than those resulting from the „Barcelona Agreement" scenario.

Commodity balances and net trade effects

The scenario of „WTO liberalisation" leads to changes in the product balances, especially for those commodities that are highly protected. For commodities that are less significant within the EU the effects are minor. The highest effects are in the livestock markets and particularly for beef, pork and poultry, where the supply in each of the EU-15 regions (i.e. Greece, Italy, Spain and rest of EU-15) decreases by 7, 4 and 11 percentage points, respectively. In the 2004 accession states the effects are milder, with decreases of 4, 1 and 4 percentage points in these three respective markets respectively.

Changes in the production of Mediterranean commodities are smaller. The supply of tomatoes rises by 1 to 2 percentage points due to „WTO liberalisation", while the production of oranges decreases by 2 to 5 percentage points. The effects on the apple market are positive for Greece and the ten 2004 accession countries (increase in the supply by 4 and 3 percentage points respectively) but negative for other EU regions, although the decrease is within a one percentage point margin. The supply changes in the cotton, tobacco and olive oil markets are minor and are close to zero in the EU-25 region. Supply falls by less than 5 percentage points in Bulgaria and Romania. This is to be expected since the EU has already opened these markets and the only distortion to free trade are domestic direct subsidies within the EU. The decrease in cotton production in Bulgaria and Romania is due to price adjustments and cross-price effects.

For other commodities there is an increase in the supply of rice by 6 percentage points and an increase of wheat production in Greece, by 3 percentage points. In both scenarios the milk and sugar markets the production quotas are not abolished but continue to be maintained. This means that there are no changes in the supply of milk and sugar throughout the EU markets.

Consumption by contrast seems to increase due to „WTO liberalisation", with the highest increases in the EU-15 region being in poultry (14 percentage points), beef (11 percentage points) and sugar (8 percentage points). The increase in the demand for poultry, beef meat and sugar is higher still in the ten new Member States (20, 34 and 18 percentage points respectively) and in Bulgaria and Romania (25, 10 and 13 percentage points respectively). In the EU-15 region the demand for rice declines by 10 percentage points, that for olive oil by 1 percentage point and for tobacco by 2 percentage points.

The changes brought about by SC5 only exist in the markets where deviations in the NPR have been modelled, i.e. in those markets where EU border measures were applied in the new Base Run scenario (SC2) or where the initial NPR of the EU was greater than zero and are twice as high as for SC4.

The adjustments on the net trade of certain agricultural markets are illustrated in Figures 5.8 and 5.9. Under the „WTO liberalisation" scenario a net trade Spain's poultry meat market, moves from being where a net exporter of almost 1,000 t p.a. to a net importer of about 266,000 t p.a., Greece moves from being a net exporter of sugar (of 16,000 t in the new Base Run Greece) to a net importer (of 11,000 t sugar) and Bulgaria and Romania change from being net exporters of wheat (21,000 t under SC2) to net importers (of 145,000 t) under SC4.

Figure 5.8: Net trade effects of „WTO liberalisation" on Mediterranean EU Member States (SC4)

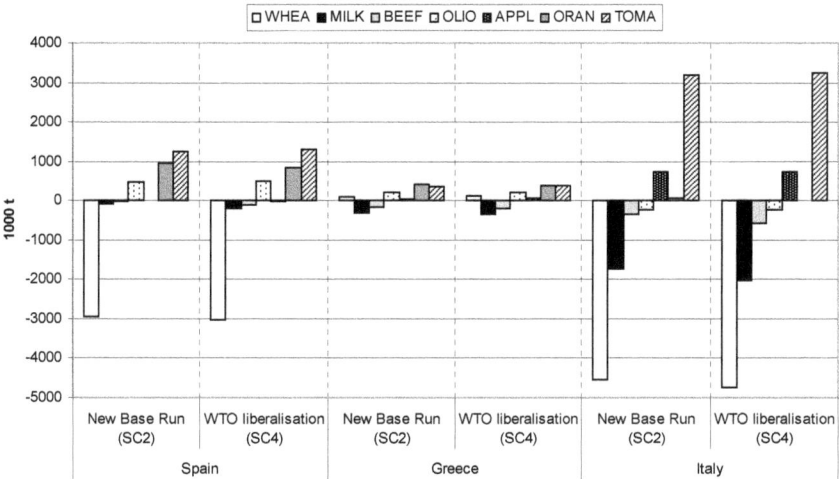

Notes: Positive values in the Y axis refer to net exports and negative values to net imports
Source: Own compilation based on AGRISIM simulations

Full liberalisation (SC5) would additionally result in a change of the net trade status of Spanish pork (from net exports of almost 0.4 million t under SC2 to net imports of 56,000 t), of the net trade status of Italian oranges, with Italy moving from being a net exporter of 70,000 t of oranges to becoming a net importer of 61,000 t, of the net trade status for sugar among the ten 2004 accession States (from net exports of 0.2 million t to net imports of 0.25 million t) and of the net trade in milk of Bulgaria and Romania, formerly net exporters of about 1.4 million t milk they become net importers of 0.6 million t.

Finally it is worth noting that the net traded quantity of olive oil in the full liberalisation scenario (SC5) is much closer to the traded quantity under the Barcelona Agreement scenario (SC3) than under WTO liberalisation scenario (SC4).

Figure 5.9: Net trade effects of „WTO liberalisation" on non-Mediterranean EU regions (SC4)

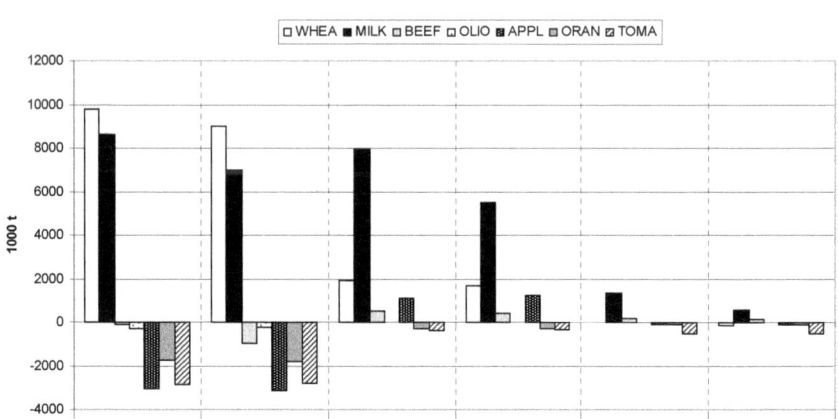

Notes: Positive values in the Y axis refer to net exports, negative values to net imports
Source: Own compilation based on AGRISIM simulations

Prices

The price effects are attributed to the simulated changes in the NPR, as shown in Table 5.17. To remind the reader, the „WTO liberalisation" scenario involved a reduction in the NPR of the EU of 50 %, while full liberalisation was simulated by eliminating the difference between domestic and world market prices, i.e. by applying a NPR of zero. Other protection measures such as production quotas or minimum farm gate prices were not altered and have remained at the same level as in the new Base Run scenario. These changes create observable deviations in both in farm gate and EU border prices, the later of which are adjusted to the level of the world market prices. The deviations are higher in markets where the changes of the NPR are the highest.

„WTO liberalisation" results in a decrease in the farm gate prices for sugar, milk, beef and poultry meat by 15 percentage points in each of Greece, Italy, Spain and rest of EU-15. In the ten new Member States the decrease in the farm gate price of beef is 21 percentage points and in Bulgaria and Romania this declines by 59 percentage points. Further declines are observed in the farm gate prices of pork and cereals, with the exception of rice. The decline is of 6 and 4 percentage points in the EU-25 countries and of 4 and 9 percentage points in Bulgaria and Romania. Although the NPR for rice is reduced, farm gate prices for this commodity rise by 22 percentage points, mainly due to cross-price effects. The farm gate prices for oranges decline by between 4 and 6 percentage points and the prices for apples by nearly 1 percentage point in all EU regions, with the only observable increase being in Greece, where there is an increase of 4 percentage points. In the rest of the markets, where no change of the NPR takes place, farm gate prices increase but only minimally.

Table 5.17: Net protection rates in the EU-25 markets, in %

Commodity	Base Year	BA	SC1	SC2	SC3	SC4	SC5
WHEA	2	2	2	2	2	1	0
COAR	8	8	8	8	8	4	0
RICE	40	40	40	40	40	20	0
OILS	0	0	0	0	0	0	0
SUGA	75	75	75	75	75	38	0
MILK	60	60	60	60	60	30	0
BEEF	164	142	142	144	150	82	0
PORK	23	23	23	23	23	11	0
POUL	49	49	49	49	49	24	0
COTT	0	0	0	0	0	0	0
TOBA	0	0	0	0	0	0	0
OLIO	0	0	0	0	0	0	0
APPL	7	7	7	7	7	3	0
ORAN	15	15	15	15	15	7	0
TOMA	0	0	0	0	0	0	0

Source: Own calculations with AGRISIM

The „WTO liberalisation" scenario leads to an increase in border prices throughout the EU-27 for nearly all markets with the exception of cereals where the border prices decrease slightly, by 3 percentage points for wheat and 1 for coarse grains. The highest increases in border prices are for beef, by 15 percentage points in the EU-25 and 8 percentage points in Bulgaria and Romania, for sugar by 8 percentage points in all EU-15 and 14 percentage points in the 2004 accession countries. In Bulgaria and Romania, the increase in border price of sugar is 85 percentage points. A further notable increase in border prices is that for milk by 6 percentage points in all the EU-27.

Farmers' revenue adjusts to these new levels of supply and farm gate prices. Thus, under SC4 cereal farmers in the EU are faced with a decline in revenue, of 5 percentage points or less. Producers of sugar, milk, beef and poultry are faced with reductions in their revenue of 15, 14, 20 and 25 percentage points respectively. Orange producers will also suffer from a revenue decrease of 6 percentage points (9 percentage points in Bulgaria and Romania), while apple producers face a decrease of nearly 2 percentage points (although in Greece they will enjoy a revenue increase of

9 percentage points). For Mediterranean commodities, such as tomatoes, olive oil, cotton and tobacco, farmers in Spain and the rest of the EU-15, are expected to enjoy a revenue increase of 2 percentage points, those in Greece an increase of 3 percentage points, those in Italy a change of between 1 to 5 percentage points, those in the ten 2004 accession states an increase of nearly 1 percentage point and those in Bulgaria and Romania an increase of 1 percentage point (apart from cotton where farmers' revenue decreases by 4 percentage points).

The price and revenue effects resulting from the full liberalisation scenario (SC5) are almost twice the magnitude of the results of SC4.

Budgetary, allocative and welfare Effects

The budgetary effects are mainly due to changes in customs duties. The changes in agricultural subsidies are due to changes in direct and input subsidies and are of a rather small magnitude, as shown in Tables 5.18 to 5.21. The changes of the subsidies are endogenous adjustments due to changes in the commoditiy balances.

Table 5.18: Changes in direct subsidies in the EU-27 due to „WTO liberalisation" (SC4) (deviations from new Base Run (SC2) in US$ million)

Commodities	Greece	Italy	Spain	rest of EU-15	new Member States (EU-10)	Bulgaria & Romania
WHEA	0	0	0	0	0	0
COAR	0	0	0	0	0	0
RICE	0	-4	-3	-1	0	0
OILS	0	0	0	0	0	0
SUGA	0	0	0	0	0	0
MILK	0	0	0	0	0	0
BEEF	0	0	0	0	0	0
PORK	0	0	1	4	0	0
POUL	0	1	1	4	0	0
COTT	0	0	0	0	0	0
TOBA	0	0	0	0	0	0
OLIO	-1	0	-1	0	0	0
APPL	0	0	0	0	0	0
ORAN	0	0	0	0	0	0
TOMA	0	0	0	0	0	0

Source: Own calculations with AGRISIM

Table 5.19: Changes in direct subsidies in the EU-27 due to full liberalisation (SC5) (deviations from new Base Run (SC2) in US$ million)

Commodities	Greece	Italy	Spain	rest of EU-15	new Member States (EU-10)	Bulgaria & Romania
WHEA	0	0	0	0	0	0
COAR	0	0	0	0	0	0
RICE	0	-1	-1	0	0	0
OILS	0	0	0	0	0	0
SUGA	0	0	0	0	0	0
MILK	0	0	0	0	0	0
BEEF	0	0	0	0	0	0
PORK	0	1	2	7	1	0
POUL	0	1	1	8	1	0
COTT	-1	0	0	0	0	0
TOBA	0	0	0	0	0	0
OLIO	-2	-1	-2	0	0	0
APPL	0	0	0	0	0	0
ORAN	0	0	0	0	0	0
TOMA	0	0	0	0	0	0

Source: Own calculations with AGRISIM

Table 5.20: Changes in input subsidies in the EU-27 due to „WTO liberalisation" (SC4) (deviations from new Base Run (SC2) in US$ million)

Commodities	Greece	Italy	Spain	rest of EU-15	new Member States (EU-10)	Bulgaria & Romania
WHEA	0	0	0	1	0	0
COAR	0	1	0	4	0	0
RICE	0	-1	0	0	0	0
OILS	0	0	0	0	0	0
SUGA	0	0	0	0	0	0
MILK	0	0	0	0	0	0
BEEF	1	11	6	50	1	0
PORK	0	1	4	12	1	0
POUL	0	4	3	21	3	0
COTT	0	0	0	0	0	0
TOBA	0	0	0	0	0	0
OLIO	0	0	0	0	0	0
APPL	0	0	0	0	0	0
ORAN	0	0	0	0	0	0
TOMA	0	0	0	0	0	0

Source: Own calculations with AGRISIM

Table 5.21: Changes in input subsidies in the EU-27 due to full multilateral trade liberalisation (SC5) (deviations from new Base Run (SC2) in US$ million)

Commodities	Greece	Italy	Spain	rest of EU-15	new Member States (EU-10)	Bulgaria & Romania
WHEA	-1	0	0	0	0	0
COAR	0	2	0	6	0	0
RICE	0	0	0	0	0	0
OILS	0	0	0	0	0	0
SUGA	0	0	0	0	1	0
MILK	0	0	0	0	0	0
BEEF	2	34	19	153	3	0
PORK	0	3	7	23	2	0
POUL	1	7	6	41	5	0
COTT	0	0	0	0	0	0
TOBA	0	0	0	0	0	0
OLIO	0	0	0	0	0	0
APPL	0	0	0	0	0	0
ORAN	0	0	0	0	0	0
TOMA	0	0	0	0	0	0

Source: Own calculations with AGRISIM

The changes in customs duties are highest in the beef and milk markets, followed by the pork market. This is due to the high levels of changes in net trade, the reduction of the NPR and chain effects on border prices. However, the effects of full multilateral liberalisation are not linear and double those as under SC4, since the customs duties are influenced by multiplying the price difference (i.e. the difference between the farm gate and the border prices) with the net trade quantity (Tables 5.22 and 5.23).

Tables 5.24 and 5.25 show the welfare effects on producers, quota owners, consumers and taxpayers in the EU-27 regions.

Table 5.22: Changes in customs duties in the EU-27 due to „WTO liberalisation" (SC4) (deviations from new Base Run (SC2)) in US$ million)

Commodities	Greece	Italy	Spain	rest of EU-15	new Member States (EU-10)	Bulgaria & Romania
WHEA	0	-5	-4	13	3	0
COAR	-4	-5	-15	3	15	4
RICE	-1	-24	-10	76	0	0
OILS	0	0	0	0	0	0
SUGA	3	0	1	255	13	-2
MILK	-12	-71	4	525	570	106
BEEF	-42	20	67	592	415	188
PORK	-36	-99	80	372	15	34
POUL	-3	56	43	407	99	-14
COTT	0	0	0	0	0	0
TOBA	0	0	0	0	0	0
OLIO	0	0	0	0	0	0
APPL	0	12	0	-61	5	-2
ORAN	9	4	37	-750	-8	-4
TOMA	0	0	0	0	0	0

Source: Own calculations with AGRISIM

Table 5.23: Changes in customs duties in the EU-27 due to full multilateral trade liberalisation (SC5) (deviations from new Base Run (SC2) in US$ million)

Commodities	Greece	Italy	Spain	rest of EU-15	new Member States (EU-10)	Bulgaria & Romania
WHEA	0	-11	-7	24	5	0
COAR	-8	-17	-36	-7	25	7
RICE	0	0	0	0	0	0
OILS	0	0	0	0	0	0
SUGA	2	-17	-12	442	15	-9
MILK	-32	-186	-8	923	900	136
BEEF	-182	-382	-19	-88	741	345
PORK	-77	-230	102	596	-25	54
POUL	-21	24	0	313	71	-81
COTT	0	0	0	0	0	0
TOBA	0	0	0	0	0	0
OLIO	0	0	0	0	0	0
APPL	0	23	0	-129	10	-3
ORAN	17	4	68	-1551	-16	-7
TOMA	0	0	0	0	0	0

Source: Own calculations with AGRISIM

Table 5.24: Welfare effects in the EU-27 due to „WTO liberalisation" (SC4) (deviations from new Base (SC2) in US$ million)

	Spain	Greece	Italy	rest of EU-15	new Member States (EU-10)	Bulgaria & Romania
producer surplus	-449	-18	-376	-3143	-1120	-392
quota owner surplus	-336	-77	-597	-4832	-741	-166
consumer surplus	892	219	1338	8555	1266	401
budget^1	213	-87	-99	1527	1133	310
total^1	320	36	266	2108	538	153
budget^2	24	137	489	394	-1043	–
total^2	130	260	854	975	-1638	–

Notes: ^{1} Without considering intra-community financial flows; ^{2} Considering intra-community financial flows
Source: own calculations with AGRISIM

Table 5.25: Welfare effects on the EU-27 due to full multilateral trade liberalisation (SC5) (deviations from new Base Run (SC2) in US$ million)

	Spain	Greece	Italy	rest of EU-15	new Member States (EU-10)	Bulgaria & Romania
producer surplus	-949	-29	-826	-6223	-2154	-755
quota owner surplus	-728	-164	1324	-10383	-1515	-342
consumer surplus	2051	528	3191	19185	2749	865
budget^1	121	-300	-746	760	1738	442
total^1	495	35	294	3339	818	210
budget^2	18	329	974	365	-1686	–
total^2	391	664	2014	2945	-2606	–

Notes: ^{1} Without considering intra-community financial flows; ^{2} Considering intra-community financial flows
Source: own calculations with AGRISIM

The simulation results of multilateral trade liberalisation, either partial or full, reveal that the EU producers are the social group who are most negatively affected. Under the EU's current policy regime they benefit from high domestic prices, which motivate them to produce more. Although the effects on the prices differ from market to market, the opening of all 15 markets within the model results in lower farm gate prices and thus a reduction in producer surplus (Table 5.25). By contrast, consumers benefit from lower prices and higher demand and enjoy welfare gains.

Simulations with AGRISIM and model results 149

The effects on the EU's budget are positive overall, but these vary between the different regions of the EU, mainly depending on each Member State's contribution to the EU budget and its net trade status. Member States that are net importers do not enjoy import levies, but pay them to the EU, whereas net exporters do not finance export subsidies from their national sources. Thus, when intra-community financial flows are taken into account the effects on taxpayers are particularly beneficial for the EU's Mediterranean Member States in comparison with the other EU regions. The budgetary effects on the ten 2004 accession states are negative, despite their low contribution to the EU budget, since the system of intra-community financial flows means that the customs duties they collect go straight into the EU budget. The picture would be different if these changes in the budget were financed solely from national resources. In this case, then the Greek and Italian budgets would be burdened with an additional US$87 and 99 million under the „WTO liberalisation" scenario and with an additional US$300 and 746 million under SC5 (Table 5.25).

The overall welfare effects show that trade liberalisation improves the social welfare in all EU regions, apart from the ten 2004 accession States, although here the negative effects are brought about by the system of intra-community financial flows. Overall, the welfare effects across the entire EU-27 are positive.

5.4.2 Effects on the MPCs

The effects of „WTO liberalisation" (SC4) on the MPCs are much smoother and of a lower magnitude than the effects of the „Barcelona Agreement". The effects of a full multilateral trade liberalisation (SC5) are twice the magnitude of those of SC4 only in those markets where border protection was previously applied.

Commodity balances and net trade effects

The changes in the commodity balances under the „WTO liberalisation" scenario are most evident in those markets that are highly protected by the MPCs. The net results are generally a decline in supply and small adjustments in demand. More detailed analysis is presented below.

In Morocco the cereal supply decreases by 2 percentage points, milk production by 6 percentage points and olive oil production by 5 percentage points. More pronounced reductions are observed in the supply of poultry meat and apples, where under SC4, production declines is by 18 and 12 percentage points respectively. By contrast the supply of tomatoes, oranges and cotton increases by 2 percentage points each.

In Turkey the results are of a small magnitude, with the highest reduction from „WTO liberalisation" being observed in supply of beef market, in the order of 5 percentage points. There is a slight increase in the supply of tomatoes and oranges, of 3 and 4 percentage points respectively, while the olive oil supply declines by 3 percentage points.

The most significant changes take place in the other MPCs, especially on the tomato market. „WTO liberalisation" leads to a decline in the supply of tomatoes of 20 percentage points, although this is still lower than the reduction under to the „Barcelona Agreement". In the rest of the markets, such as olive oil, the effects on supply are not more than -5 percentage points.

The adjustments on demand are also smoother under SC4 than under SC3 („Barcelona Agreement"). The highest changes in demand viz à viz the new Base Run (SC2) are observed on the Moroccan apple market, and the Turkish beef market, where demand increases by 8 percentage points, and on the tomato market in the other MPCs, where demand increases by 6 percentage points. In the other markets the demand effects are very small and are less than +/- 3 percentage points.

The effects of full multilateral liberalisation on supply and demand are of twice the magnitude to SC4, but only for those markets where a change in the NPR has been modelled.

The net trade effects are illustrated in Figure 5.10. The general tendency is for MPCs to slightly increase their imports but also to slightly increase exports of typical Mediterranean commodities, such as oranges or tomatoes.

Figure 5.10: Net trade effects on MPCs due to „WTO liberalisation" (SC4)

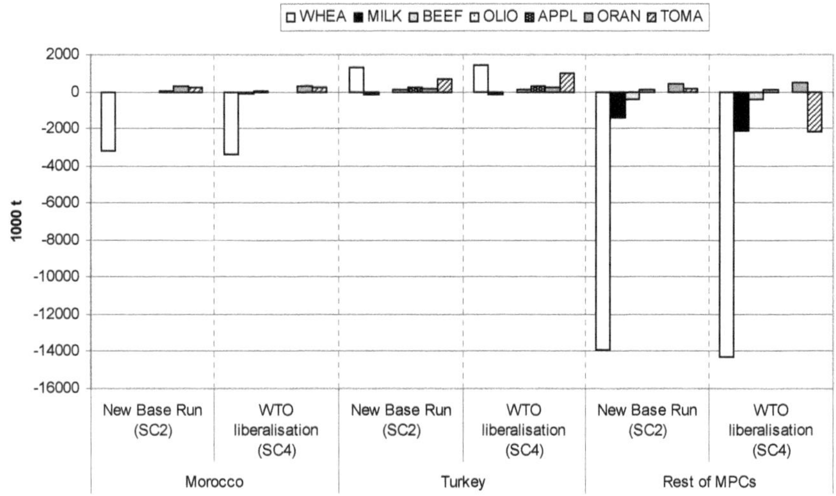

Notes: Positive values in the Y axis refer to net exports, negative values to net imports
Source: Own compilation based on AGRISIM simulations

Simulations with AGRISIM and model results 151

A striking exception to this trend is that of the tomato market in the other MPCs. Not only do exports decrease but the net trade status of the region changes. Starting as a net exporter of 0.14 million t, the region becomes a net importer of 2.16 million t under „WTO liberalisation" (SC4) and 4.5 million t under full liberalisation (SC5).

For olive oil, another typical Mediterranean commodity, the changes in the trade balance are not in favour of the MPCs, although they are of a quite small level. Exports from Turkey decrease by 4,000 and 9,000 t under SC4 and SC5 (respectively). Those from the other MPCs decrease by 30,000 and 66,000 t, respectively, under the two scenarios. Morocco sees a change in its net trade status moving from being a net exporter of 2,000 t to a net importer, of 3,000 and 8,000 t under the two respective scenarios.

Further changes of the net trade status are observed in Morocco and Turkey. Morocco's net trade status in apples changes, with the country moving from being a net exporter of 2,000 t to a net importer of 3,000 t under SC4, and of 8,000 t under SC5. Turkey's net trade status in beef changes moving from being a net exporter of almost 1,000 t under SC2 to a net importer of 42,000 and 122,000 t under SC4 and SC5 respectively. No further changes in net trade status are observed as a result of simulating a full multilateral trade liberalisation.

Prices

The reduction of the results in adjustments to farm gate prices and border prices in all MPCs (Table 5.26). The effects vary between the single markets and the countries-regions because of the different initial protection. The changes in producer incentive prices are equal to those in farm gate prices, because the model does not include any agricultural subsidies that act as a price incentive to farmers.

In Morocco the farm gate prices of rice, milk and poultry are most affected by „WTO liberalisation", all of them falling by 25 percentage points. The farm gate prices for olive oil and apples also fall, by 15 percentage points respectively, while those for wheat fall by 13 percentage points. By contrast the farm gate prices for beef, pork and sugar increase by 5 percentage points each while those for oranges and tomatoes increase by not more than 4 and 1 percentage points respectively.

The picture is different in Turkey, with SC4 leading to a decline in the farm gate prices of beef of 24 percentage points, and a 10 percentage points decline in farm gate prices for coarse grains, oilseeds and olive oil. Reductions of a lower magnitude are observed for the farm gate prices for poultry meat (7 percentage points), sugar (6 percentage points) and cotton (5 percentage points). The farm gate prices for apples, oranges and tomatoes are expected to increase by 3, 4 and 2 percentage points respectively.

Table 5.26: Net protection rates in the MPCs' markets, in %

Commodity	Morocco Base Year	SC4	SC5	Turkey Base Year	SC4	SC5	Rest of MPCs Base Year	SC4	SC5
WHEA	29	14	0	-7	-3	0	7	3	0
COAR	0	0	0	19	9	0	0	0	0
RICE	111	56	0	0	0	0	0	0	0
OILS	0	0	0	25	13	0	0	0	0
SUGA	0	0	0	27	14	0	2	1	0
MILK	115	57	0	16	8	0	26	13	0
BEEF	0	0	0	207	103	0	8	4	0
PORK	0	0	0	0	0	0	2	1	0
POUL	109	55	0	19	10	0	20	10	0
COTT	3	1	0	0	0	0	0	0	0
TOBA	18	9	0	25	12	0	4	2	0
OLIO	55	27	0	31	16	0	43	22	0
APPL	55	27	0	0	0	0	7	3	0
ORAN	0	0	0	0	0	0	7	3	0
TOMA	0	0	0	0	0	0	37	19	0

Source: Own calculations with AGRISIM

In other MPCs the most distinguishable changes in farm gate prices under the „WTO liberalisation" scenario are for olive oil, tomatoes, sugar, milk and poultry meat, which change by -14, -12, +11 and -8 percentage points respectively.

The adjustments to border prices are of lower magnitude than those to farm gate prices and some general trends are observable. Overall the border prices increase, with the highest increases being for livestock commodities (mainly beef) and sugar. In Turkey the border price for beef is by 15 percentage points higher than under the new Base Run (SC2), while in the rest of the MPCs the sugar border prices increase by 12 percentage points. For Mediterranean commodities, the increase in border prices throughout the MPCs is 2 to 4 percentage points. Only in the wheat and coarse grains markets is a decrease in border prices, of 2 percentage points.

The changes in farmers' revenue are equal to the changes in farm gate prices and supply. It is difficult to draw general conclusions on the changes in farmers' revenue since the effects vary between individual countries, however some trends can be drawn in relation to particular markets.

„WTO liberalisation" leads to a revenue decrease of 10 percentage points for poultry producers throughout the MPCs, with Moroccan producers facing a decrease in revenue of almost 40 percentage points. The revenue of olive oil growers also decreases, by between 12 percentage points in Turkey to nearly 20 percentage points in Morocco. Moroccan producers of milk, rice, apple and wheat are also faced with significant decreases in their revenue, of 30, 25 and 15 percentage points respectively. In Turkey tomato farmers face a revenue decrease of 30 percentage points, and milk producers one of 12 percentage points. By contrast Turkish sugar farmers would enjoy a revenue increase of 12 percentage points. In the other MPCs, distinguishable changes in revenue levels are found among farmers of coarse grains and oilseeds (a decrease of 11 percentage points for each commodity) among sugar farmers (a decrease of 7 percentage points) and orange and tomato farmers (an increase of the same magnitude). For the other commodities farmers' income levels o remain as in the new Base Run scenario.

The application of full multilateral liberalisation has results that follow the same general direction as those from SC4, although they are of about twice the magnitude.

Budgetary, allocative and welfare effects

The budgetary effects are attributed to changes in customs duties (Table 5.27), which are in turn the result of changes in the net traded quantities and the prices. Changes in agricultural subsidies (direct, input and general) are close to zero.

Table 5.27: Changes in customs duties in MPCs due to „WTO liberalisation" (SC4) and full multilateral trade liberalisation (SC5) (deviations from new Base Run (SC2) in US$ million)

Commodity	Morocco		Turkey		Rest of MPCs	
	SC4	SC5	SC4	SC5	SC4	SC5
WHEA	-72	-150	-5	-12	-73	-147
COAR	0	0	3	-3	0	0
RICE	0	-1	0	0	0	0
OILS	0	0	-35	-72	0	0
SUGA	0	0	28	57	-3	-7
MILK	15	-7	-3	-5	-48	-202
BEEF	0	0	37	1	-33	-67
PORK	0	0	0	0	0	0
POUL	15	-3	0	1	8	0
COTT	0	-1	0	0	0	0
TOBA	-2	-4	9	19	-7	-15
OLIO	4	2	25	48	54	85
APPL	5	2	0	0	0	-1
ORAN	0	0	0	0	4	9
TOMA	0	0	0	0	163	18

Source: Own calculations with AGRISIM

Opening of trade, either partial or full, results in a decrease of revenue from import tariffs but also in lower expenditure on export subsidies. The combination of these effects has a negative impact on Morocco's overall budget (compared to the new Base Run level) and a positive one for Turkey. In the rest of the MPCs the changes in budgetary effects stemming from SC4 and SC5 are a result of changes in the customs duties for tomatoes, which are in turn due to changes in the net trade status.

The allocation of resources is clearly in favour of consumers, a result that was expected and which reflects the theoretical considerations of the benefits of trade liberalisation (as discussed in chapter 3). Under the „WTO liberalisation" scenario consumers benefit from an increase of their surplus of US$0.5 billion in Morocco, US$0.3 billion in Turkey and of US$1.6 billion in the rest of the MPCs. Producers, however, are worse off, with their surplus decreasing by US$0.5, 0.2 and 1.4 billion in the three regions respectively. The deviations of the effects of SC5 are about twice the magnitude, as shown in detail in Table 5.28.

Table 5.28: **Welfare effects on MPCs due to „WTO liberalisation" (SC4) and full multilateral trade liberalisation (SC5) (deviations from SC2 in US$ million)**

	Morocco		Turkey		Rest of MPCs	
	SC4	SC5	SC4	SC5	SC4	SC5
producer surplus	-457	-859	-227	-387	-1476	-2625
quota owner surplus	0	0	0	0	0	0
consumer surplus	524	1028	283	552	1608	3024
budget	-36	-162	60	37	63	-329
total	31	7	116	202	195	69

Source: Own calculations with AGRISIM

The relatively small magnitude of the results in terms of overall welfare and the negative budgetary effects on Morocco and the rest of the MPCs compare favourably with the findings of earlier studies, such as that of BUSSE and GROIZARD (2007) and indicates that openness to trade in itself is not a sufficient condition to provide gains. Other factors seem to play an important role, such as for example geographical variables or institutional quality. This argument is also supported by the findings of BORRMANN and BUSSE (2006a),[20] which imply that countries with good government regulations are more likely to benefit from trade. Certainly it would worth examining whether the MPCs require institutional reforms so as to support an efficient market structure and a well functioning allocation of resources among the producers and the

---

[20] A shorter version of this study has been published as BORRMANN and BUSSE (2006b)

consumers. Such a conclusion is supported by the other empirical studies analysing the effects of trade liberalisation on MPCs cited in chapter four.

5.5 The competitiveness of Mediterranean Countries

The issue of competitiveness is a complex one that can include many aspects. One indication of where one country is placed, compared to the rest of the world can be drawn by observing changes in its trade balance and especially the changes in the trade balance of the main import and export commodities, using these as an approximation of changes in the shares in commodities that a country has in other markets. Another indication is changes in a country's production, compared to the rest of the world, since the ability to maintain its supply share can indicate low production costs, high yields and the existence of markets where these commodities can be sold, which together imply a degree of competitive in supplying a given commodity.

With this in mind the competitiveness of Mediterranean countries, both the EU Member States and the MPCS, is examined by looking changes in their trade balances and changes in the world supply (i.e. their share of production) of the main export and import commodities. Because of the high competition between the MPCs and the EU Mediterranean Member States, the focus here is upon the changes due to the forthcoming Mediterranean Free Trade Area, a possible extension of this into a customs union (SC3) and a possible outcome of the WTO negotiations (SC4). Comparisons are made with the new Base Run (SC2), which is a close representation of the present day status quo.

As mentioned in section 5.2.1, the changes in net imports and net exports between SC2 and SC3 are rather marginal for the EU Mediterranean countries (Greece, Italy and Spain). The net trade status quo of these countries does not change and they remain net importers of "northern" commodities, such as wheat, beef and milk, with the exception of Greece, which is a net exporter of 95,000 t wheat under both SC2 and SC3. They remain as net exporters of typical Mediterranean commodities, such as tomatoes, oranges, olive oil, cotton and tobacco (with the exception of Italy, which is the largest net importer of olive oil among the EU-27). The „Barcelona Agreement" brings only very small changes to the quantities traded and there is a general trend of trade performance developing in the favour of these three countries.

However, the effects of „WTO liberalisation" (SC4) on the southern EU countries are of a higher magnitude, as presented in section 5.3.1. In this scenario the trade performance of the three EU Mediterranean Member States only develops positively in respect of the tomato market. In general their imports increase and their exports decrease, with the most affected commodities being oranges, milk and beef. This in turn leads to some changes in their trade status, as for example in the case of the Greek sugar market.

Figure 5.11 illustrates the deviations between the present situation (SC2) and the „Barcelona Agreement" (SC3) and „WTO liberalisation" (SC4) scenarios.

The picture differs slightly when looking at the net trade effects on the other side of the Mediterranean basin, i.e. on the MPCs (section 5.2.2). Under SC3, the trade balance of Morocco, Turkey and the rest of the MPCs seems to deteriorate slightly for markets that are not supported in the EU (i.e. the EU's NPR is close to zero). For example their net imports of cereals increase and their net exports of olive oil decrease. By contrast the effects are positive for markets for which the EU levies a high protection rate, such as beef and milk. In these markets the trade balance develops in favour of the MPCs. Trade in tomatoes is affected in different ways by the „Barcelona Agreement" with Turkey and Morocco exporting more, and the rest of the MPCs not only exporting less, but also moving from net exporters to become net importers.

Figure 5.11: Net trade effects of the „Barcelona Agreement" (SC3) and „WTO liberalisation" on Spain, Greece and Italy (SC4)

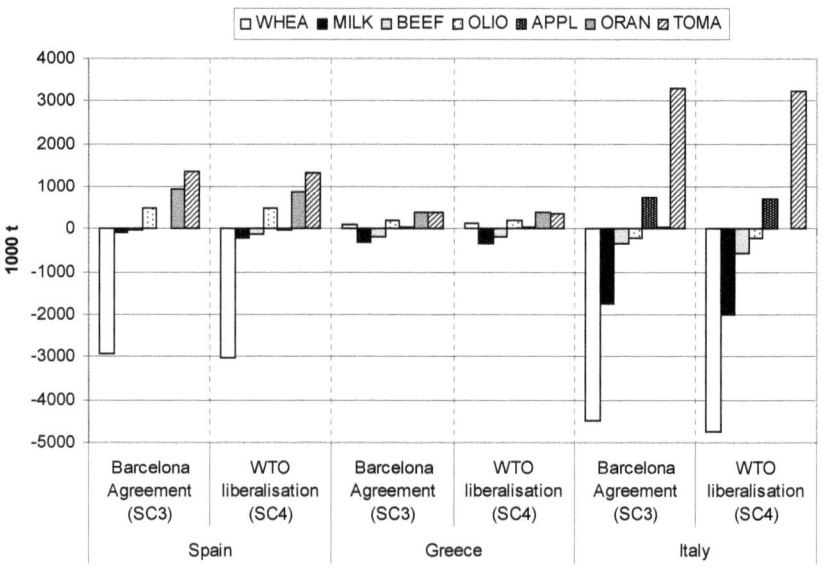

Notes: Positive values in the Y axis refer to net exports, negative values to net imports
Source: Own compilation based on AGRISIM simulations

The effects of „WTO liberalisation" (SC4) on the MPCs are milder and changes of a lower magnitude, as presented in detail in section 5.3.2. Compared to the new Base Run (SC2) their imports of cereals slightly increase, but they decrease compared to the „Barcelona Agreement" (SC3), as Figure 5.12 shows. Exports of typical Mediterranean commodities, such as olive oil, decrease slightly compared to the new Base Run, but by less than under the „Barcelona Agreement". This same pattern holds for the trade balance of tomatoes among the rest of MPCs, where the „WTO liberalisation" leads to a decrease in exports and increase in imports, which is lower than under the „Barcelona Agreement". The same applies for the increase in net exports of oranges.

Figure 5.12: Net trade effects of the „Barcelona Agreement" (SC3) and „WTO liberalisation" (SC4) on Morocco, Turkey and the other MPCs

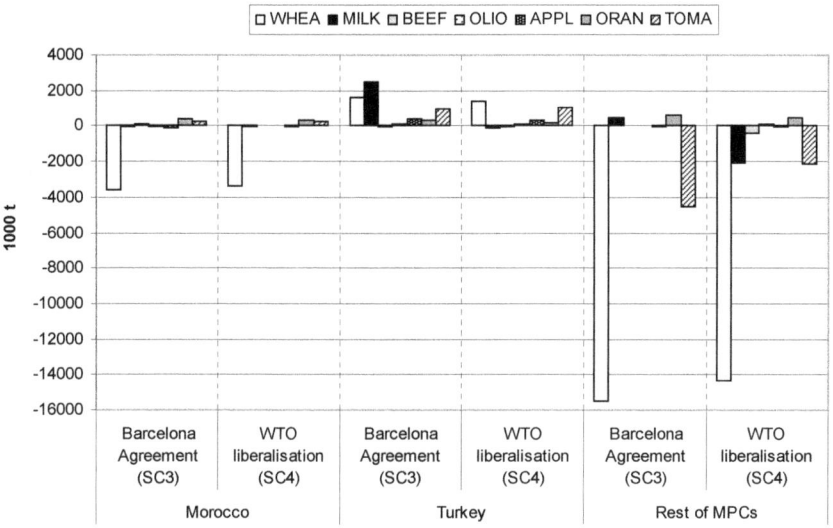

Notes: Positive values in the Y axis refer to net exports, negative values to net imports
Source: Own compilation based on AGRISIM simulations

Despite the changes in net trade status, the changes in the shares of world production and supply are only marginal. The „Barcelona Agreement" and „WTO liberalisation" have virtually no effects on total world supply and the Mediterranean countries, as a whole, maintain their original market shares.

The EU southern Member States are by far the world's largest supplier of olive oil (producing about 80 % of the world's olive oil). SC3 allows them to increase their share of world production by 1.0 percentage poing. Under the „WTO liberalisation" scenario this increase is about half at about +0.5 percentage points. This group of countries supplies about 9 %, 11 %, 2.6 % and 5 % of the world's supply of oranges, tomatoes, cotton and tobacco respectively. The „Barcelona Agreement" and „WTO liberalisation" scenarios only give rise to changes in his initial market of less than +/-

0.5 percentage points. Figure 5.13 gives an overview of the changes in these three countries' shares in the world production of Mediterranean commodities and of apples.

Figure 5.13: Changes in the world supply shares of EU Mediterranean Member States (in %)

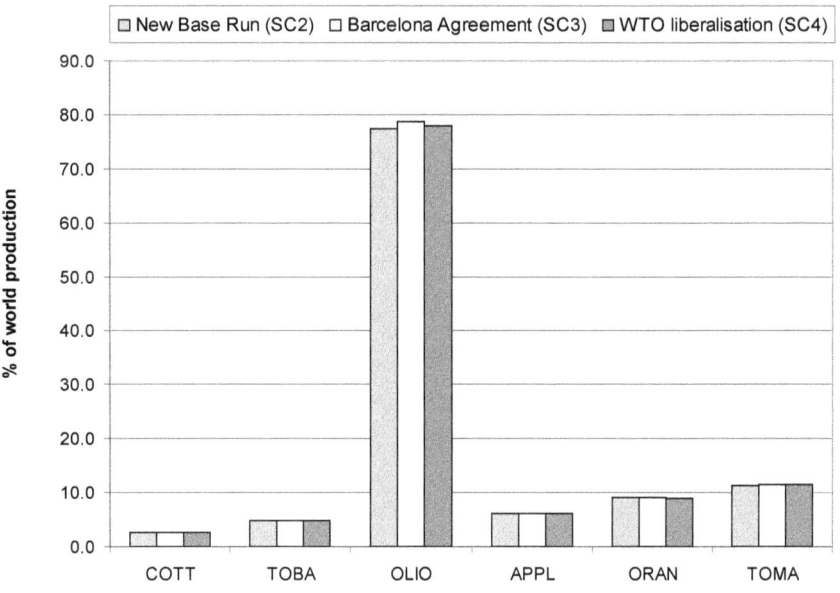

Source: Own compilation based on AGRISIM simulations

In addition the simulated scenarios show that these three countries maintain their share of production within the EU-27, for all these commodities, compared both to the Base Run (BA) and to the new Base Run (SC2) - as shown in detail in Table B.8. For example throughout the simulations Greece remains the most important cotton producer in the EU and Spain the biggest olive oil producer. 82 % of the EU's cotton production and 58 % of the EU's olive oil production take place in Greece and Spain respectively.

The MPCs also maintain their shares of world production, although these were rather small, less than 5 % for all the commodities included in the model, except for olive oil and tomatoes, where they account for 18 % of the world's olive oil and tomato production. Throughout the simulations the deviations from the base year and among the scenarios are only minor. Distinguishable deviations are observed in relation to tomatoes, where the „Barcelona Agreement" leads to an approximate 4 percentage point decrease in the rest of the MPCs share of production and the „WTO liberalisa-

Simulations with AGRISIM and model results 159

tion" leads to fall of 2 percentage points. The scenario changes also have small effects these countries market share for olive oil, which declines by 0.2 percentage points in Morocco and Turkey and by 1.0 percentage point in the rest of MPCs. For Turkey's supply share of tomatoes increases by 0.4 percentage points while that of the rest of the MPCs decreases by 4.0 points (Table B.7 and Figure 5.14). It should be also noted that, apart from olive oil, where the EU's southern Member States supply most of the world's output, the EU Mediterranean Member States and the MPCs have broadly comparable shares of world production for other typical Mediterranean commodities.

Figure 5.14: Changes in world supply shares of all MPCs (in %)

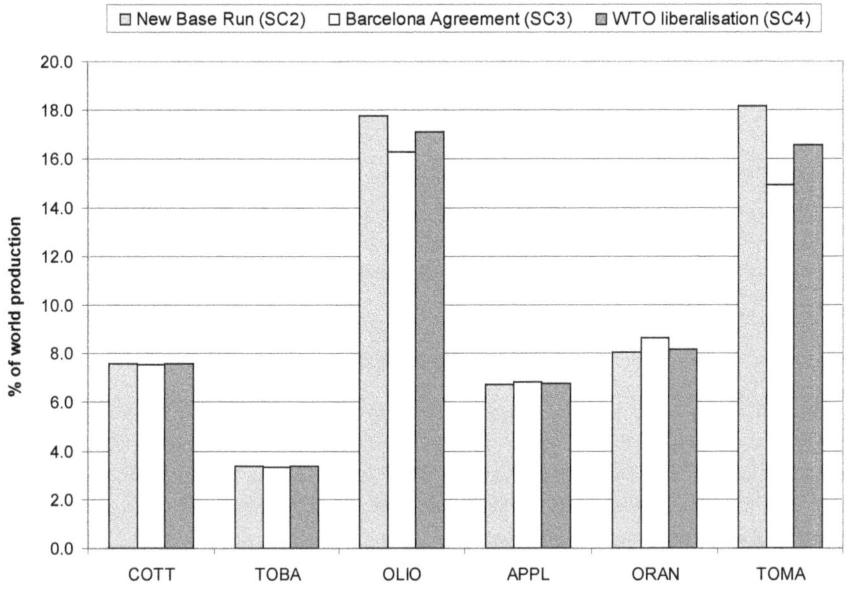

Source: Own compilation based on AGRISIM simulations

Figure 5.15 illustrates the changes in world supply between the new Base Run (SC2), the „Barcelona Agreement" (SC3) and „WTO liberalisation" (SC4). The deviations in absolute numbers are all within 3 percentage points, with the largest deviations being in the MPCs' share of tomato supply under the „Barcelona Agreement" scenario.

Figure 5.15: Changes in world supply shares of typical Mediterranean commodities and of apples (deviations from new Base Run (SC2) in percentage points)

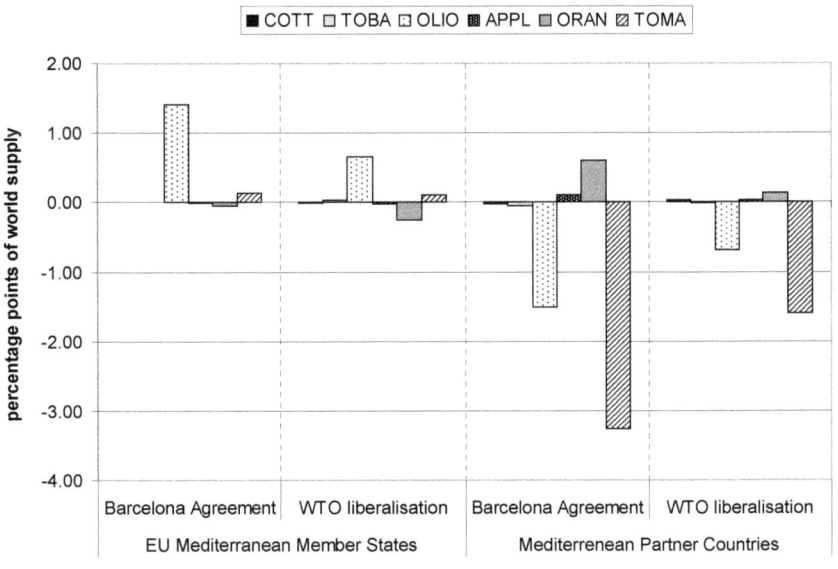

Source: Own compilation based on AGRISIM simulations

From the above it can be concluded that the Free Trade Area and the potential customs union between the EU and the MPCs will not affect significantly the relative competitiveness of the countries involved. The Mediterranean EU Member States not only maintain their relative competitiveness, but marginally improve their market access for the commodities that they export.

These three countries' ability to maintain their initial production shares and the slight improvement of their exports under the „Barcelona Agreement" scenario suggest that these countries will not be discriminated against the other EU regions by the Euro-Med Agreements. Therefore, this analysis should allay the fears of Greece, Italy and Spain that they will lose market shares for their commodities as a result of increased competition from the MPCs in the wake of the Barcelona Agreement.

The MPCs by contrast do experience a slight deterioration of their relative competitiveness under the third and fourth simulated scenarios („Barcelona Agreement" and „WTO liberalisation" respectively), although the effects are only marginal. This outcome could justify some of the scepticism that the MPCs show towards the Euro-Med Agreements.

6 Concluding remarks

Agricultural trade around the Mediterranean basin has been based on traditional relationships between Mediterranean countries. In recent years several attempts have been made to boost it through numerous trade agreements between the EU and non-EU Mediterranean Partner Countries (MPCs). Despite the fact that the MPCs are involved in several regional trade agreements, which stem back from the '80's, regional integration has remained low. A new page for integration around the Mediterranean started with the EU's initiative to complete the Euro-Med Agreements through the Barcelona Process.

The forthcoming creation of a Free Trade Area between the EU and non-EU Mediterranean countries within the framework of the Barcelona Agreement and changes to the EU's agricultural policy due to the ongoing WTO negotiations and domestic policy reforms offer new opportunities and new challenges for agriculture within Mediterranean countries. The lack of empirical studies on the effects of such likely policy changes on the agricultural sector around the Mediterranean and especially their effects on main traded commodities from the region creates the need for analysing the likely impacts of policy reforms on the trade and welfare of the countries involved.

This study examines the reforms to agricultural policy that are likely to occur within the Mediterranean basin and their likely effects on agricultural markets in the EU and the MPCs. Simulations of various liberalisation options and intended reforms to the EU's agricultural policy reveal the effects on commodity balances, net trade, prices, budgets, welfare and competitiveness within these blocks of countries. Emphasis is given to the regional effects on the Mediterranean countries, i.e. those within EU, particularly Greece, Italy and Spain and the MPCs. The simulations have been carried out with an extended and modified version of the partial equilibrium trade policy model, AGRISIM.

The simulation results vary from scenario to scenario, from country to country (and/or region to region) and from market to market within specific regions, thus making it difficult to formulate conclusions on general trends.

The CAP reform of Mediterranean commodities, i.e. the decoupling of direct payments for cotton, olive oil and tobacco (and especially the 50 % cut in direct payments for tobacco) seem to only slightly affect countries within the EU that produce these commodities. The impacts on other EU regions and on the MPCs are also minimal. The small magnitude of the results can be explained by the fact that decoupling only changes the method of supporting farmers, and does not result in any further structural reforms. Equally these markets are small and of limited importance to both the EU-27 and the MPCs.

The main effects of the most recent EU enlargement to include Bulgaria and Romania are mostly felt only by these two countries, and have almost no impact on the agricultural markets of other EU regions or of the MPCs.

The observations from these first two scenarios were used to formulate a new Base Run scenario, which includes both the CAP reform for cotton, olive oil and tobacco and the enlargement of the EU to include Bulgaria and Romania.

The analysis of a potential extension of a Free Trade Area between the EU and the MPCs into a customs union (SC3) shows only a marginal effect on the commodity balances and prices within the EU regions. The deviations of this scenario from the new Base Run scenario (SC2) show that the trade performance of EU Mediterranean Member States seem to improve slightly, as they are able to export slightly more and to import slightly less. They will also enjoy welfare gains. However, the welfare effects on the other EU regions are negative, due to reduced customs levies.

The effects of the Barcelona Agreement on the MPCs vary, depending on the changes in border protection. Overall their trade performance for their main export commodities, such as olive oil and oranges, deteriorates slightly and they suffer from some welfare losses. The simulations in this scenario involve an extreme situation with the creation of both a Free Trade Area and a customs union between the EU and the MPCs, taking the scenario to the upper limits of the Euro-Med Agreements. The results show that the EU Mediterranean countries will not be worse off from a full implementation of the Barcelona Agreement and will potentially benefit from an expansion of trade. Moreover, agricultural producers in these countries will not be negatively affected. However, the effects of the scenario on the MPCs are negative and justify the scepticism that these countries have shown towards the Barcelona process.

The scenario of „WTO liberalisation" has more marked effects on EU markets, but less effect on the markets of the MPCs. In general, the domestic supply for commodities that are highly protected in the EU, such as beef, declines, while demand increases. The net trade performance of Greece, Italy and Spain only develops positively for tomatoes. In both the EU regions and the MPC regions, producer surplus declines and consumer surplus increases. The effects on the EU's budget are positive, revealing overall welfare gains for the EU-27, while the effect on the budget of the MPCs is negative, due to a reduction in import revenues. In the event of full trade liberalisation the effects are twice the magnitude, but only in those markets where border protection previously existed.

The effects of SC5 on the MPCs are closer to the results of SC3, than those from SC4.

Mediterranean countries, both within and outside the EU maintain their supply shares and there are only marginal effects on their net trade, implying that their relative competitiveness is not affected by these simulations.

To go back to the fourth question in the introduction, over the budgetary effects in the EU and in specific Member States, the simulations showed these to be positive, since the reduction of trade distorting policies implies lower expenditure. As the CAP is financed according to the principle of financial solidarity and member states' individual contributions are based on their Gross National Input, the budgetary effects differ between the regions of the EU. The reform of the CAP, the enlargement of the EU, the implementation of the Euro-Med Agreements and the opening of EU markets is beneficial for the taxpayers of member states that make a small contribution to the EU budget, such as Greece, Italy and Spain. By contrast taxpayers in the northern EU member states and the new member states experience negative effects, the former due to the high contribution they make to the common budget and the later because of losing import rents that they previously enjoyed. Under the new Base Run Scenario, the changes in the budget caused by the „Barcelona Agreement" and „WTO liberalisation" favour taxpayers in the EU-15. Without the intra-community financing system the effects on the EU Mediterranean member states would be of a lower magnitude, positive under SC3, but negative for Greece and Italy under SC4, implying that the EU financing system is particularly beneficial for Greece, Italy and Spain.

The third question in the introduction concerned the effects of these changes on producers and consumers. From the changes in consumer surplus it is clear that consumers in both the EU and the MPCs enjoy welfare gains under multilateral liberalisation, but are worse off in the event of a customs union between the EU and the MPCs. By contrast producers in both sets of countries are worse off due under multilateral liberalisation, but those in the EU benefit from the implementation of the Barcelona Agreement (SC3), with those in the MPCs being worse off.

Thus the fears of farmers in southern EU countries that their incomes and welfare will suffer from the joint effects of the CAP reform for Mediterranean products and the Barcelona Agreement are not confirmed by the model results, although a multilateral opening of EU markets may have negative consequences for them. Under the current protected policy regime producers enjoy higher prices than they would under a liberalised policy regime. Some adjustments would be required to offset the negative effects of a liberalised policy regime on farmers' welfare. These might include policies that encourage efficiency in production, logistics and marketing so as to maximise the benefits from a well functioning supply chain.

There is little difference in the changes in farmers' revenue in the north and south of the EU, showing that the scenarios explored do not unduly discriminate against farmers in EU Mediterranean states. However, it should be noted that the magnitude of the effects of these changes upon typical Mediterranean commodities, such as cotton, olive oil and tobacco, is much higher if the producer's incentive prices are used as the basis for calculating farmers' revenue, largely because of a reduction in direct payments made to them.

The second question in the introduction, asked about the combined effects on EU farmers and the regional budgetary effects within the EU. Here the analysis reveals that not only are the southern EU Member States not discriminated against by the agricultural policy reforms (in comparison to other EU regions), but that they benefit more than other EU regions, due to the intra-EU system of financial flows. In addition they show a potential for improving their trade balance, even in the event of a customs union with the MPCs, whose producers are their main competitors.

This leads back to the first question in the introduction, about the effects of trade liberalisation on Mediterranean agriculture. Under the scenarios Mediterranean agriculture does not appear to be worse off, but also not significantly better off. In comparison with the rest of the world, the Mediterranean countries maintain approximately the same degree of competitiveness under multilateral trade liberalisation. Equally a Free Trade Area within the Mediterranean does not seem to change the relative competitiveness of the EU Mediterranean Member States and non-EU Mediterranean countries, although the effects on individual countries are differentiated, as discussed previously.

One point which clearly emerged from the simulations, which should be telling for policy makers is that trade liberalisation, whether multilateral (within the framework of the WTO) or bilateral (within the framework of the Barcelona Agreement), leads to overall welfare gains and that the effects on markets are of a small magnitude. The results of this study compare favourably with those from the empirical studies cited in the fourth chapter. Therefore moving towards liberalisation would not harm markets but, on the contrary, would be beneficial. The small magnitude of the effects can be attributed to the fact that the policies simulated here only refer to the level of applied protection and thus only examine the impact of differences between domestic and world market prices. The effects of non-tariff barriers or quality issues are not addressed. Recently a discussion has emerged over such policies, particularly quality issues that relate to imported commodities, such as sanitary and phytosanitary criteria and other non-tariff barriers. These policies could, in the long run, result in a shift in the supply and demand curves and their impacts could be greater than the impacts of changes in price differences. One challenge for future research would be to quantify such policies, in terms of how much and in which direction they shift the curves.

This leads to the conclusion that policy reforms which sooner or later are inevitable, such as the abolition of tariffs, will perhaps be less of an influence on the agricultural sector around the Mediterranean than structural changes and non-tariff barriers. Structural changes are crucial for Mediterranean agriculture, since an efficiently structured system is a pre-requisite for the re-allocation of resources (initially within the sector and subsequently within the overall economy) that will allow compensation of any losses incurred by farmers. The MPCs require a proper infrastructure that can promote effective production structures and an efficient distribution of resources. This is a necessary condition for overcoming any possible negative impacts of non-tariff barriers, such as food quality criteria for exports to the EU.

The establishment of common protocols of origin could represent a way to overcome most of the existing non-tariff barriers relating to food quality. Although this process has started, with the Pan Euro Mediterranean Protocol of Origin, the numerous bilateral trade agreements, which embody different protocols of origin, complicates its adoption. Adoption of common rules of origin within these various trade agreements among the Mediterranean countries is problematic, since it involves costs in terms of establishing a new system and the necessary administration, as discussed in the second chapter. The situation is complicated by the fact that, until now, all the initiatives for regional integration and the required administrative support have been dominated by the EU and its institutions. The MPCs have not taking an active role in participating, but have been quite passive observers. This situation reinforces the argument made above, that structural changes are needed in the MPCs and that good (and active) governance is of prime importance for the future of the agricultural sector in these countries.

One limitation of the study is the comparative-static nature of the model. Although through shift factors and the possibility to model population growth do allow some dynamic aspects to be captured, the results must be seen as comparative-static. The model is not suited for prognosis and the results should be interpreted more as possible trends rather than as absolute predictions. This said, the model is suitable for a "with and without" policy analysis. Moreover, because of its comparative-static nature, the model underestimates the true gains from trade and liberalisation, since these gains are commonly much higher in reality than the comparative-static results. This is because trade liberalisation is a dynamic rather than a static procedure and the positive welfare effects are often of a higher magnitude than the comparative-static analysis suggests.

Further issues that were not examined in this study, but remain worth investigating are the effects of further regional agreements, such as the Cotonou Agreement which established an Economic Partnership Agreement (EPA) between the EU and the African, Caribbean and Pacific Countries (ACPs) and the impacts of the EU's trade negotiations with the Mercosur States.

Another area worthy of further investigation is the regional impacts within the countries concerned, particularly the impacts on rural areas within the MPCs. Since there are differences in the effects between consumers and producers, these may have regional implications, which need to be further investigated. The future development of the economies of MPCs often relies on the development of their rural areas, which in turn is highly dependent on agricultural activities. There is a concern that while trade liberalisation brings aggregate benefits the regional distribution of these within countries, especially between less and more favoured regions may be negative.

7 Summary

Bilateral and multilateral trade agreements have been gaining attention and relevance in recent years. After the deadlock in the Doha Negotiations in the summer of 2006, international attention has focused on regional trade integration and particularly on the initiatives of the European Union (EU) to conclude or deepen existing trade agreements with blocks of countries in Africa, Asia and the Pacific.

For the non-EU countries around the Mediterranean the most important of the numerous regional trade agreements in which they are involved are the Euro-Med Agreements, which influence their trade flows and seek to promote deeper regional integration.

At present trade flows of agricultural commodities around the Mediterranean are along on a north-south axis, with trade focused on specific commodities. The northern EU Member States export cereals, meat and dairy products to the MPCs, while in turn the MPCs mainly export fruits, vegetables and olive oil to the northern EU countries. This places them in direct competition with the EU Mediterranean member states. The high similarity in the exports of the MPCs and the EU Mediterranean member states and the high complementarity in trade among the MPCs and northern EU member states has created diverging interests within the EU (between northern and southern member states) which have very different attitudes towards the Euro-Med Agreements and any strategy towards further agricultural trade liberalisation.

Trade preferences within the Euro-Med Agreements have been granted so far only from the EU to the MPCs. The value of the preference margin for selected agricultural commodities reveals that the benefits for the MPCs have been slightly intensified because of the Barcelona Agreement but the gains are different from country to country and are only for fruits and vegetables. The MPCs on the other side have not granted preferences to the EU and keep high protection rates for those commodities that they import mainly from the EU as the nominal protection rate shows, such as livestock commodities and cereals,. This seems to contradict their willingness to deepen their trade relationships with the EU and benefit from this preferential treatment.

Trade among the MPCs remains rather low, despite these countries being connected with each other through a number of trade agreements. Similarities in production and exports certainly play a key role in low south-south integration. Exports destined to the EU and trade with the EU appear more attractive and leave little space for intra-south trade.

The forthcoming creation of a Free Trade Area between the EU and the MPCs, foreseen by the Barcelona Agreement, the partial trade liberalisation that might result from the ongoing WTO negotiations and domestic policy reforms (i.e. of the CAP) offer new opportunities and new challenges for the agricultural sector around the

Mediterranean basin, but also make the trade relationships round the Mediterranean to look more complex and create uncertainty to the farmers, which in turn creates the need for further analysis.

Based on the theory of applied welfare economics, this study empirically shows that the latest CAP reform will lead to only small adjustments in producer surplus and will only have a limited effect on the well being of producers. The magnitude of these effects largely depends upon how much the new decoupled subsidies affect producers' decision making processes, which will reflect the impact that the existing direct subsidies have on production.

Theoretical assessments of Free Trade Areas suggest that the welfare effects are negative for third countries and are clearly positive for the partner countries, when the partner countries that are net suppliers are more efficient (or at least as efficient) as third country suppliers. The magnitude of these effects depends upon several factors: the existing set of tariffs, the relative amount of imports from preferential and non-preferential suppliers, expenditure on domestic goods and the level of substitution between domestically produced and imported goods. Because of the interplay of these factors, it is not clear whether the trade creation effects are greater or lesser than the trade diversion effects. Thus there is some ambiguity about the effects of the Euro-Mediterranean Free Trade Area on social welfare. This implies the need for further empirical analysis. However, it is clear that preferential suppliers enjoy the preference margin granted by partner countries that are net importers.

Multilateral liberalisation is generally beneficial for welfare but not always for countries that are already involved in a regional trade agreement and enjoy preference margins. This is connected to the issue of preference erosion. A theoretical assessment of preference erosion shows positive welfare effects for third countries and negative ones for preferential suppliers, especially if they are less efficient than third countries. The effects on preferential importers are neither clearly positive nor negative. This analysis is highly relevant for the Mediterranean Partner Countries due to the regional trade agreements that exist between them and with the EU.

Existing empirical assessments of trade liberalisation around the Mediterranean basin have mostly been undertaken using General Equilibrium Models, focusing on the effects of the abolition of import tariffs between the EU and the MPCs on the secondary and tertiary sectors (i.e. manufactured goods and services). The studies have usually focused on individual MPCs and most studies show disparities between the EU and the examined (non–EU) Mediterranean country. Few empirical studies focus on the agricultural sector, and especially on the main agricultural commodities produced in Mediterranean countries. Those that exist either do not take into account simultaneous policy changes (i.e. reform of the EU's agricultural policy and trade liberalisation) or do not assess the regional disparities between farmers in different EU regions. Thus there was a gap in the analysis of the impacts of policy reforms on trade and the welfare of the involved countries, which this study addresses.

The main objective of this study has been to provide deeper insights into the impacts of agricultural and trade policy reforms in the Mediterranean basin. Further objectives include evaluating the effects of these new policy regimes, suggesting further changes to them and providing sound empirical results that can inform the discussions of policy makers over the future of Mediterranean agriculture.

In particular the thesis seeks to address the following questions:

1. What are the likely overall impacts of trade liberalisation for the agricultural sectors in the EU and the MPCs?
2. What are the likely regional effects within the EU? Do the policy changes discriminate against southern EU Member States in comparison to the northern ones?
3. What are the likely effects on producers and consumers within the two main trading blocks (i.e. the EU and MPCs)?
4. What are the likely budgetary effects for the EU as a whole and for particular Member States?

This study empirically analyses the impacts of different policy scenarios on Mediterranean countries by using an extended and modified version of the trade policy model, AGRISIM.

AGRISIM is an acronym of AGricultural SImulations Model and is a synthetic simulation model, comparative-static and deterministic in nature, with non linear, iso elastic demand and supply functions. It is a net trade model that assumes homogenous products. Regions are connected with each other through a market clearing mechanism, in which the derived world market price feeds into domestic markets through domestic prices. The net aggregate trade from all regions, which is given by the difference between supply and demand, is fed back into the world market clearing mechanism. It considers the effects of different policy interventions, such as changes in the nominal protection rate, price transmission elasticities, minimum producer prices, production quotas and subsidies. Through shift coefficients in the demand and supply functions additional variables, such as population and income growth, can be simulated. The model is used to simulate the effects on production, consumption, trade, domestic prices, border prices, state budgets, consumer and producer surpluses and overall welfare.

The model was updated to the year 2001 and extended to include all the Mediterranean Partner Countries and five commodities (apples, cotton, olive oil, oranges, tobacco and tomatoes) so as to cover the main markets of Mediterranean countries. Particularly for the EU Member States a new module has been programmed enabling to take into account the intra-community financial system of the EU when calculating the budget effects that entail in each of the EU regions.

The simulations covered four policy scenarios:

a) Further Common Agricultural Policy (CAP) reform
b) EU enlargement
c) WTO liberalisation
d) Preferential Treatment of non-EU Mediterranean countries (MPCs)

These policy scenarios were addressed through five simulation scenarios that model the differences against a Base Run scenario.

The Base Run scenario simulates the full implementation of Agenda 2000, the 2004 eastern enlargement of the EU and the Luxembourg Agreement. SC1 simulates the effects of the CAP reform of three Mediterranean commodities: cotton, olive oil and tobacco. SC2 simulates the enlargement of the EU to include Bulgaria and Romania. These two scenarios are carried forward into the other scenarios. Because the results of these scenarios are of a very small magnitude and because together they represent the current policy regime, they are used as the new Base Run.

SC3 simulates the implementation of the forthcoming Free Trade Area between the EU and the Mediterranean Partner Countries. Because the model assumes homogenous commodities and does not offer the possibility of modelling bilateral trade relationships, this scenario goes one step further and simulates a customs union between the EU and the MPCs. This is accomplished by applying the EU's trade policy with the rest of the world, as expressed through the Nominal Protection Rate (NPR), to the MPCs. The results of this scenario show the upper bounds of the full implementation of the Barcelona Agreement, thus this scenario is named the „Barcelona Agreement".

In SC4 and SC5 two different options for liberalisation are simulated. SC4 simulates a 50 % multilateral liberalisation, which is a feasible likely outcome of the ongoing WTO negotiations. This scenario is referred to as „WTO liberalisation". SC5 simulates a 100 % multilateral liberalisation. Although this scenario has a very limited practical use, since it is an unlikely outcome, it is useful for checking the plausibility of the model results and for understanding their relative magnitude.

An overview of the assumptions of each scenario follows in Table 7.1.

Table 7.1: Overview of simulated scenarios

	Base Run	SC1	SC2	SC3	SC4	SC5
Agenda 2000	✓	✓	✓	✓	✓	✓
2004 EU eastern enlargement	✓	✓	✓	✓	✓	✓
Luxembourg Agreement	✓	✓	✓	✓	✓	✓
CAP reform of Mediterranean commodities		✓	✓	✓	✓	✓
EU enlargement with Bulgaria and Romania			✓	✓	✓	✓
„Barcelona Agreement"				✓		
„WTO liberalisation"					✓	
Full multilateral liberalisation						✓

Source: Own compilation

The simulation results indicated that the CAP reform of Mediterranean commodities only very slightly affects the markets for cotton, olive oil and tobacco and then only within the EU Mediterranean member states. The enlargement of the EU to include Bulgaria and Romania only affects these two countries. The very small magnitude of the results from these two scenarios allowed SC2 to be re-formulated as the new Base Run scenario, providing an accurate representation of the existing policy regime. Throughout the thesis the results from SC3, SC4 and SC5 are compared with the new Base Run.

The „Barcelona Agreement" has very limited effects on all the EU regions, whether Mediterranean or non-Mediterranean. The effects on the net trade performance of the southern EU Member States are slightly positive, and there are almost no effects on the other EU regions. Changes in welfare favour producers and are slightly unfavourable to consumers, due to a slight increase in farm gate prices. The budgetary effects throughout the EU regions are slightly negative, due to a small reduction in customs duties. These effects vary between the EU regions, according to the financial flows between national budgets and that of Brussels. These results could be due the EU's trade with the MPCs being small compared to intra-EU trade and to its trade with the rest of the world, implying that a customs union with the MPCs would have little or no effect on EU markets. It also shows that the EU Mediterranean Member States should not fear losing their market shares or seeing a decline in their agricultural sectors due to higher competition from the MPCs.

The effects of the „Barcelona Agreement" on the MPCs vary, depending on the changes in levels of border protection for each market. Overall a slight deterioration in their trade performance is observed for the main exported commodities, such as olive oil and oranges. Equally these countries will suffer from welfare losses, due to a reduction in the trade levies that they currently collect. However, producers will benefit from higher farm gate prices and thus enjoy an increase in their surplus. In total the effects on the MPCs are unfavourable and can justify their scepticism towards the Barcelona process.

The „WTO liberalisation" scenario has more marked effects on EU markets. In general it leads to a reduction in the internal supply of highly protected commodities, such as beef, while the demand for these increases. The net trade performance of Greece, Italy and Spain only develops positively for tomatoes. As expected from the theoretical analysis of multilateral market liberalisation, the producer surplus declines and the consumer surplus increases, in all EU regions. The effects on the EU's budget are positive, revealing overall welfare gains for the EU-27. In the event of full trade liberalisation the effects are twice the magnitude, but only for those markets border protection previously existed.

The „WTO liberalisation" scenario has less effect on the markets of the MPCs than SC3. Their net trade performance declines somewhat but the changes are of a very small magnitude, except for markets that were initially high protected. There is a reduction in farm gate prices, which leads to a decrease in producer surplus and farmers' revenue, but also to an increase consumers' wellbeing as measured by an increase in consumer surplus. Only in Morocco does this scenario have a negative effect on the budget, due to a reduction in import revenues. The effects of SC5 are twice the magnitude in markets where border protection initially (in the new Base Run scenario) was applied. In terms of welfare SC5 results in positive budget changes for Turkey (but not the other MPCs). Interestingly the effects of this scenario on the MPCs are closer to the results of the Barcelona Agreement (SC3) than those from SC4.

None of the simulations affect the competitiveness of the Mediterranean countries, either within or outside the EU.

From the model results it is clear that trade liberalisation, either multilateral (within the framework of the WTO) or bilateral (within the framework of the Barcelona Agreement) leads to overall welfare gains, with the effects on markets being of a small magnitude. This compares favourably with the outcome of the empirical studies cited in the fourth chapter of this thesis. It seems that liberalisation does not leave the Mediterranean agriculture any worse off, but also not significantly better off. This point, that liberalisation of the markets would not be harmful, but on the contrary it would be beneficial for agricultural markets due to the positive welfare effects should be taken into consideration by policy makers.

The small magnitude of the effects on agricultural markets might be attributed to the relatively minor importance of the MPCs as trade partners with the EU or to the simulations that were run, which mostly capture shocks in the level of applied protection. The scenarios have focused simulating various differences between domestic and world market prices, expressed through different levels of NPR. Policies relating to non-tariff barriers or quality issues have not been examined. Discussion about the

Summary

effects of such policies has recently emerged, particularly over the quality of imported commodities, such as the sanitary and phytosanitary criteria and/or other non-tariff barriers. In the long run these policies could result in shifts in the supply and demand curves, the impacts of which might exceed those caused by any changes in price. One challenge for future research would be to seek to quantify the extent and direction of such shifts, through a quantitative modelling exercise.

In respect of the discussion about regional disparities within the EU, it seems that the fears of farmers in southern EU countries that the reform of the CAP for Mediterranean products and the Barcelona Agreement will reduce their income and have negative impacts on them are unfounded. EU farmers are only worse off under a multilateral opening of EU markets. The analysis shows an improvement in the net trade performance of the EU Mediterranean member states, with an increase in their trade balance, following a customs union with the MPCs, their main competitors. The changes in farmers' revenue in northern and southern EU member states are quite similar, indicating that the EU's Mediterranean farmers are not unduly discriminated against (compared to their colleagues in other EU regions) by these changes in agricultural policy. However, the magnitude of the effects is higher for typical Mediterranean commodities, such as cotton, olive oil and tobacco when the producer's incentive prices are used to calculate farmers' revenue, due to modelling of direct payments. The regional budgetary effects in the EU reveal that the Mediterranean EU Member States benefit more from the intra-EU financial flows between national budgets and that of Brussels' budget. This evidence also contradicts the argument that the EU's agricultural policy discriminates against the southern EU regions.

At risk of generalisation, since the results do vary from market to market, Tables 7.2 to 7.5 give a compact overview of the results of the model for the EU regions and the MPCs.

Table 7.2: Evaluation of policy scenarios on EU Mediterranean Member States

simulated scenarios	changes of:						
	net trade	farm gate prices	world supply shares	producer surplus	consumer surplus	budget	overall welfare
new Base Run (SC2)	0	0	0	-	0/-	-	-
„Barcelona Agreement" (SC3)	0/+	+	0	+	-	+	+
„WTO liberalisation" (SC4)	-	-	0	-	+	+	+
full multilateral liberalisation (SC5)	-	-	0	-	+	+	+

Notes: +: positive; –: negative; 0: almost no change; 0/-: slightly negative; 0/+: slightly positive

For the budgetary effects and overall welfare, intra-EU financial flows have been taken into account.

Source: Own compilation based on AGRISIM simulations

Table 7.3: Evaluation of policy scenarios on northern EU Member States

simulated scenarios	changes of:						
	net trade	farm gate prices	world supply shares	producer surplus	consumer surplus	budget	overall welfare
new Base Run (SC2)	0	0	0	-	+	+	+
„Barcelona Agreement" (SC3)	0	+	0	+	+	+	0/-
„WTO liberalisation" (SC4)	0	-	0	-	+	+	+
full multilateral liberalisation (SC5)	0/-	-	0	-	+	+	+

Notes: +: positive; –: negative; 0: almost no change; 0/-: slightly negative; 0/+: slightly positive

For the budgetary effects and overall welfare, intra-EU financial flows have been taken into account.

Source: Own compilation based on AGRISIM simulations

Summary

Table 7.4: Evaluation of policy scenarios on new EU Member States (EU-10)

simulated scenarios	changes of						
	net trade	farm gate prices	world supply shares	producer surplus	consumer surplus	budget	overall welfare
new Base Run (SC2)	0	0	0	-	+	0/-	-
„Barcelona Agreement" (SC3)	0	+	0	-	+	-	-
„WTO liberalisation" (SC4)	0/-	-	0	-	+	-	-
full multilateral liberalisation (SC5)	0/-	-	0	-	+	-	-

Notes: +: positive; –: negative; 0: almost no change; 0/-: slightly negative; 0/+: slightly positive
For the budgetary effects and overall welfare, intra-EU financial flows have been taken into account.

Source: Own compilation based on AGRISIM simulations

Table 7.5: Evaluation of policy scenarios on the MPCs

simulated scenarios	changes of						
	net trade	farm gate prices	world supply shares	producer surplus	consumer surplus	budget	overall welfare
new Base Run (SC2)	0	0	0	0	0	0	0
„Barcelona Agreement" (SC3)	0/-	+	0	+	-	-	-
„WTO liberalisation" (SC4)	0/-	-	0	-	+	+	+
full multilateral liberalisation (SC5)	0/-	-	0	-	+	-	+

Notes: +: positive; –: negative; 0: almost no change; 0/-: slightly negative; 0/+: slightly positive

Source: Own compilation based on AGRISIM simulations

List of References

AGHROUT, A. (2007): Embracing Free Trade: The EU's Economic Partnership with Algeria. Intereconomics 42 (2): 96-104.

ALESSANDRI, P. (2000): European and Euro-Mediterranean Agreements: Some Simulation Analysis on the Effects of the EU Trade Policy. Working Paper No. 10, Centro Studi sui Processi di Internazionalizzazione (CESPRI), Milano.

ALEXANDRAKI, A. and H. P. LANKDES (2004): The Impact of Preference Erosion on Middle-Income Developing Countries. Working Paper WP/04/169, International Monetary Fund (IMF), Washington D.C.

AMAD (various years): Agricultural Market Access Database. Available from: http://www.amad.org/pages/0,2966,en_35049325_35049378_1_1_1_1_1,00.html, accessed at 10/2003-10/2005.

ANDERSON, J. E. and E. van WINCOOP (2003): Gravity with Gravitas: A Solution to the Border Puzzle. The American Economic Review 93 (1): 170-192.

AQUILA, C. DELL' and B. E. VELAZQUEZ (2002): Euromed Agreements and Agricultural Trade Issues. Working Paper No. 16, Italian National Institute of Agricultural Economics (INEA), Rome.

ARFINI, F., DONATI, M. and D. MENOZZI (2005): Analysis of the Socio-Economic Impact of the Tobacco CMO Reform on Italian Tobacco Sector. Poster background paper prepared for presentation at the XIth Congress of the European Association of Agricultural Economists (EAAE), Copenhagen, August 24-27.

ARFINI, F. (2001): Mathematical Programming Models Employed in the Analysis of the Common Agricultural Policy. Working paper No. 9, Osservatorio sulle Politiche Agricole dell' UE, Italian National Institute of Agricultural Economics (INEA), Rome.

ARMINGTON, P. S. (1969): A Theory of Demand for Products Distinguished by Place of Production. IMF Staff Papers No. 16, International Monetary Fund (IMF), Washington D. C.

AUGIER, P. and M. GASIOREK (2003a): The Welfare Implications of Trade Liberalisation between the Southern Mediterranean and the EU. Applied Economics 35 (10): 1171-1190.

―――― (2003b): Partial Equilibrium Modelling of the EU-Med Agreements: The Case of Morocco. Contributed paper at the Femise Conference on Euro-Mediterranean Association Agreements and the EU's New Neighbourhood Policy Vision, Marseilles, December 4-6.

BALKHAUSEN, O., BANSE, M. and H. GRETHE (2008): Modelling CAP Decoupling in the EU: A Comparison of Selected Simulation Models and Results. Journal of Agricultural Economics 59 (1): 57-71.

BALKHAUSEN, O., BANSE, M., GRETHE, H. and S. NOLTE (2005): Modelling the Effects of Partial Decoupling on Crop and Fodder Area as well as Ruminant Supply in the EU: Current State and Outlook. In: ARFINI, F. (ed.): Modelling Agricultural Policies: State of the Art and New Challenges. Proceedings of the 89th European Seminar of the European Association of Agricultural Economists (EAAE). Parma: Monte Università Parma, 565-587.

BANSE, M., GRETHE, H. and S. NOLTE (2004): European Simulation Model (ESIM) in GAMS: Model Documentation. Model Documentation prepared for DG AGRI, European Commission, Göttingen and Berlin.

BAUER, S. and H. KASNAKOGLU (1990): Non-linear Programming Models for Sector and Policy Analysis: Experiences with the Turkish Agricultural Sector Model. Economic Modelling 7 (3): 275-289.

BAYAR, A. H., BEN-AHMED, G., DIAO, X., and A. E. YELDAN (2001): An Inter-Temporal, Multi-Region General Equilibrium Model of Agricultural Trade Liberalisation in the South Mediterranean NIC's, Turkey and the EU. Second Conference du Femise, Institut de la Mediteranée, Marseille, March 29-30.

BAYAR, A. H., DIAO, X. S., and A. E. YELDAN (2000): An Inter-Temporal, Multi-Region General Equilibrium Model of Agricultural Trade Liberalisation in the South Mediterranean NIC's, Turkey and the EU. Discussion Paper No. 6, Trade and Macroeconomics Division (TMD), International Food Policy Research Institute (IFPRI), Washington D.C.

BERGSTRAND, J. H. (1985): The Gravity Equation in International Trade: Some Microeconomic Foundations and Empirical Evidence. The Review of Economics and Statistics 67 (3): 474-481.

BHAGWATI, J. (1999): Regionalism and Multilateralism: An Overview. In: BHAGWATI, J., KRISHNA, P. and A. PANAGARIYA (eds.): Trading Blocs: Alternative Approaches to Analysing Preferential Trade Agreements. Cambridge and London: The MIT Press, 3-32.

BHAGWATI, J. and A. PANAGARIYA (1996): Preferential Trading Areas and Multilateralism – Strangers, Friends or Foes? In: BHAGWATI, J. and A. PANAGARIYA (eds.): The Economics of Preferential Trade Agreements. Washington D. C.: The AEI Press, 1-78.

BINFIELD J., MEYERS, W. and P. WESTHOFF (2005): Challenges of Incorporating EU Enlargement and CAP Reform in the GOLD Model Framework. In: ARFINI, F. (ed.): Modelling Agricultural Policies: State of the Art and New Challenges. Proceedings of the 89th European Seminar of the European Association of Agricultural Economists (EAAE). Parma: Monte Università Parma, 291-306.

BINFIELD J., DONNELLAN T., HANRAHAN K. and P. WESTHOFF (2003): The Luxembourg CAP Reform Agreement: Implications for EU and Irish Agriculture. In: FAPRI-Ireland Partnership (ed.): The Luxembourg CAP Reform Agreement: Analysis of the Impact on EU and Irish Agriculture. Teagasc Rural Economic Centre. Available from: http://tnet.teagasc.ie/fapri/downloads/pubs2003/luxag/fullreport141003.pdf, accessed at 5/03/2004.

List of references

BORRMANN, A. and M. BUSSE (2006a): The Institutional challenge of the ACP/EU Economic Partnership Agreements. Research Paper No. 2-3, Hamburg Institute of International Economics (HWWI), Hamburg.

BORRMANN, A. and M. BUSSE (2006b): Institutional Prerequisites of Economic-Partnership Agreements: The ECOWAS Case. Intereconomics 41 (4): 231-236.

BOUËT, A. (2006a): What can the Poor Expect from Trade Liberalisation? Opening the „Black Box" of Trade Modelling. Markets. Discussion Paper No. 93, Trade and Institutions Division (MTID), International Food Policy Research Institute (IFPRI), Washington D.C.

—— (2006b): Defining a Trade Strategy for Southern Mediterranean Countries. Discussion Paper No. 97, Markets, Trade and Institutions Division (MTID), International Food Policy Research Institute (IFPRI), Washington D.C.

BOUZERGAN, A. (2007): The European Union and its ten Mediterranean Partner Countries: Growing Trading Links. Statistics in Focus, External Trade No. 70/ 2007, Eurostat, European Communities, Luxembourg.

BREEN, J. P., HENNESSY, T. C. and F. S. THORNE (2005): The Effect of Decoupling on the Decision to Produce: An Irish Case Study. Food Policy 30 (2): 129-144.

BRITZ, W. (2005): CAPRI Modelling System Documentation: Common Agricultural Policy Impact Analysis. Development of a Regionalised EU-Wide Operational Model to assess the Impact of Current Common Agricultural Policy on Farming Sustainability, J05/30/2004 – Deliverable 1, Bonn.

BRITZ, W., JUNKER, F. and L. WEISSLEDER (2006): Quantitative Assessment of the EU-Mediterranean Trade Liberalisation using the CAPRI Modelling System. Deliverable 24 prepared for the Project SSPE-CT-2004-502457 „EU-Med AgPol: Impacts of Agricultural Trade Liberalisation between the EU and the Mediterranean Countries", October (revised in November). Available from: http://eumedagpol.iamm.fr/html/ publications/ prj_report/d24.pdf, accessed at 10/02/2007.

BROCKMEIER, M. (1999): Die Relevanz allgemeiner Gleichgewichtsmodelle für die agrarökonomische Forschung. Agrarwirtschaft 48 (2): 438-447.

BROCKMEIER M., HEROK, C. A., LEDEBUR, O. VON and P. SALAMON (2003): EU Enlargement-A New Dimension. Paper prepared for presentation at the 25th International Conference of Agricultural Economists, Durban, August 16-22.

BROWN, D. K., DEARDORFF, A. V. and R. M. STERN (1997): Some Economic Effects of the Free Trade Agreement between Tunisia and the European Union. In: GALAL, A. and B. HOEKMAN (eds.): Regional Partners in Global Markets: Limits and Possibilities of the Euro-Med Agreements. London: Centre for Economic Policy Reasearch (CEPR), Cairo: The Egyptian Center for Economic Studies (ECES), 71-97.

BURFISHER, M. E., SHERMAN, R. and K. THIERFELDER (2003): Regionalism: Old and New, Theory and Practise. Invited paper presented at the International Agricultural Trade Research Consortium (IATRC), Capri, June 23-26.

BUSSE, M. and J. L. GROIZARD (2007): Does Africa really benefit from Trade?. Research Paper No. 2-7, Hamburg Institute of International Economics (HWWI), Hamburg.

BUTCHER, A. M., BALISTERI, E., CHRIST, N., FOX, A., JOHNSON, K., TSIGAS, M., WILSON, E., HONNOLD, V., DENNIS, A., JOHNSON, C., MCCARTY, T., REEDER, J., RODRIQUEZ, L., STELLER, R., BROWN, R., BLOODGOOD, L., NUNE, M., GEARHART, W. W., RIMMER, D., SUMMERS, J. L. and P. AUGUSTINE (2004): U.S.-Morocco Free Trade Agreement: Potential Economywide and Selected Sectoral Effects. Investigation No. TA-2104-14, Publication 3704, United States International Trade Commission (USITC), Washington D.C.

BUTCHER, A., CHOMO, G. V., RAPKINS, D., AUSTIN, N., BAKER, S., BURKET, S., CANTRELL, R., DENNIS, A., FREUND, K., MCCARTY, T., PAULSON, M., REEDER, J., RORKE, J., STELLER, R., WANSER, S., FERRANTINO, M., HALL, K. and P. POGANY (2000): Economic Impact on the United States of a U.S.-Jordan Free Trade Agreement. Investigation No. 332-418, Publication 3340, United States International Trade Commission (USITC), Washington D.C.

CAHILL, S. A. (1997): Calculating the Rate of Decoupling for Crops under CAP/ Oilseeds Reform. Journal of Agricultural Economics 48 (3): 349-378.

CAKMAK, E. H. and H. KASNAKOGLU (2003): The Impact of EU Membership on Agriculture in Turkey. Contributed paper at the Femise Conference on Euro-Mediterranean Association Agreements and the EU's New Neighbourhood Policy Vision, Marseilles, December 4 – 6.

CHATTI, R. (2003): A CGE Assessment of FTA between Tunisia and the EU under Oligopolistic Market Structures. Review of Middle East Economics and Finance 1 (2): 99-127.

CHEMINGUI, M. A. and S. DESSUS (2001): The Liberalisation of Tunisian Agriculture and the European Union: A Prospective Analysis. In: DESSUS, S., DEVLIN, J. and R. SAFADI (eds.). Towards Arab and Euro-Med Regional Integration. Paris: OECD, 147-168.

Commission of the European Communities (2007a): Proposal for a Council Regulation laying down specific Rules as regards the Fruit and Vegetable Sector and amending certain Regulations. SEC(2007) 74, SEC(2007) 75, COM(2007) final, 2007/0012 (CNS), Brussels.

—— (2007b): Commission Staff Working Document accompanying Document to the Proposal for a Council Regulation laying down specific Rules as regards the Fruit and Vegetable Sector and amending certain Regulations, Impact Analysis Summary. SEC(2007) 74, COM(2007) 17 final, Brussels.

—— (2004): CAP Reform – Accomplishing Sustainable Agriculture: the Tobacco, Olive Oil, Cotton, Sugar and Hops Sectors: Outline of the Commission proposals. Available from: http://europa.eu.int/comm/agriculture/capreform/docs/prop2 _ en.pdf, accessed at 21/06/2004.

List of references

—— (2003a): Proposal for a Council Regulation amending Regulation (EC) No 1782/ 2003 establishing Common Rules for Direct Support Schemes under the Common Agricultural Policy and establishing certain Support Schemes for Farmers, Proposal for a Council Regulation on the Common Organization of the Market in Olive Oil and Table Olives and amending Regulation (EEC) No 827/68. 698 final. Available from: http://europa.eu.int/comm/agriculture/ capreform/com698_en.pdf, accessed at 21/06/ 2004.

—— (2003b): Communication from the Commission to the Council and the European Parliament – Accomplishing a Sustainable agricultural Model for Europe through the reformed CAP – The Tobacco, Olive Oil and Sugar Sectors. 554 final. Available from: http://europa.eu.int/comm/agriculture/capreform/com554/554_en.pdf, accessed at 21/06/2004.

—— (2003c): Tobacco Regime – Extended Impact Assessment. Commission Staff Working Document, European Commission, European Union, Brussels.

COMTRADE (various years): Commodity and Trade Database. United Nations Statistics Division. Available from: http://comtrade.un.org/, accessed at: 05/2005-05/2006.

CONFORTI, P. (2001): The Common Agricultural Policy in Main Partial Equilibrium Models. Working Paper No. 7, Osservatorio sulle Politiche Agricole dell' UE, Italian National Institure of Agricultural Economics (INEA), Rome.

CONFORTI P. and P. LONDERO (2001): AGLINK: The OECD Partial Equilibrium Model. Working Paper No. 8, Osservatorio sulle Politiche Agricole dell' UE, Italian National Institute of Agricultural Economics (INEA), Rome.

Council of the European Union (2004): 2575th Agriculture and Fisheries Council Meeting – Luxembourg, 21-22 April 2004. Press: 112 No. 8401/04. Available from: http://ue.eu.int/ueDocs/cms_Data/docs/pressData/en/agricult/80010.pdf, accessed at 21/06/04.

CRAWFORD, J.-A. and R. V. FIORENTINO (2005): The Changing Landscape of Regional Trade Agreements. Discussion Paper No. 8, World Trade Organisation (WTO), Geneva.

DENNIS, A. (2006): The Impact of Regional Trade Agreements and Trade Facilitation in the Middle East North Africa Region. Policy Research Working Paper No. 3837, The World Bank, Washington D.C.

DIAO, X. and A. E. YELDAN (2001): An Intertemporal, Multi-Region General Equilibrium Model of Agricultural Trade Liberalization in the South Mediterranean NIC's, Turkey, and the European Union. In: HAGEN, J. VON and M. WIDGREN (eds.): Regionalism in Europe, Geometries and Strategies After 2000. Massachusetts: Kluwer Academic Publishers, 195-220.

Directorate-General for Agriculture (2003): The Cotton Sector. Working Paper, European Commission, European Union, Brussels.

ELBEHRI, A. and T. W. HERTEL (2004): A Comparative Analysis of the EU-Morocco FTA vs. Multilateral Liberalization. GTAP Working Paper No. 31, West Lafayette.

ERDLE, S. (2007): Die europäische Nachbarschaftspolitik, Ein Motor für Reformen im Mittelmeerraum? KAS/Auslandsinformationen 4/07, 6-40.

EU COMMISSION (2008a): The Euro-Mediterranean Partnership / Barcelona Process. DG External Relations. Available from: http://ec.europa.eu/external_relations/euromed/, accessed at 01/2008.

—— (2008b): Barcelona Declaration adopted at the Euro-Mediterranean Conference 27-28/11/95. DG External Relations. Available from: http://ec.europa.eu/external_relations/euromed/ bd.htm, accessed at 01/2008.

—— (2008c): The Mediterranean Region. DG Trade. Available from: http://ec.europa.eu/trade/issues/bilateral/regions/euromed/index_en.htm, accessed at 24/01/2008.

—— (2006): The Euro-Mediterranean Partnership. DG External Relations. Available from:http://europa.eu.int/comm/external_relations/euromed/med_ass_agreements.htm, accessed at 26/03/06.

EUROPEAN COMMISSION (2003): Mid-Term Review of the Common Agricultural Policy July 2002 Proposals Impact Analyses. DG Agriculture, European Union, Brussels.

EUROPEAN COMMUNITIES (2005): Financial Report 2004. European Union. Available from: http://ec.europa.eu/budget/library/publications/fin_reports/fin_report_04_en.pdf, access-ed at 12/01/2007.

EUROSTAT (2006): Euro-Mediterranean Statistics. Panorama of the European Union, General and Regional Statistics, European Commission, Luxembourg.

—— (various years): COMEXT (Commission of External trade) Database. Available from http://fd.comext.eurostat.cec.eu.int/xtweb/, accessed at 2005-2007.

FAO (2005): The COSIMO Work Programme at FAO. Committee on Commodity Problems, Sixty-Fifth Session, Food and Agriculture Organisation of the United Nations (FAO), Rome, April 11-13. Available from: http://www.fao.org/docrep/meeting/009/ J4756e.htm, accessed at 06/06/2006.

FAOSTAT (various years): Agricultural Statistical Database. Food and Agriculture Organisation of the United Nations (FAO). Available from: http://faostat.fao.org/, accessed at 2003-2006.

FAPRI (various years): Elasticities Database. Food and Agricultural Policy Research Institute (FAPRI). Available from: http://www.fapri.iastate.edu/tools/elasticity.aspx, accessed at 10/2003-03/2006.

FERABOLI, O. (2006): Preferential Trade Liberalisation and Public Finance Reform: A Dynamic CGE Model for Jordan. Paper presented at the 10th Conference on Theories and Methods in Macroeconomics, Université des Sciences Sociales, Toulouse-Greqam, January 19-20.

FERABOLI, O., GAITAN, B. and B. LUCKE (2003): Trade Liberalisation and the Euro-Med Partnership: A Dynamic Model for Jordan. Diskussionsarbeit, Institut für Wachstum und Konjuktur, Universität Hamburg.

FAA (2002): Modellanalyse zu den Auswirkungen der KOM-Vorschläge im Rahmen der Halbzeitbewertung der Agenda 2000. Schriftenreihe der Forschungs-gesellschaft für Agrarpolitik und Agrarsoziologie (FAA), Bonn.

FRANCOIS, J. F., HOEKMAN, B. and M. MANCHIM (2006): Preference Erosion and Multilateral Trade Liberalisation. The World Bank Economic Review 20 (2): 197-216.

FRANCOIS, J. F. and H. K. HALL (1997): Partial Equilibrium Modelling. In: FRANCOIS, J. F. and K. A. REINERT (eds.): Applied Methods for Trade Policy Analysis. Cambridge: Cambridge University Press, 122-155.

FRANCOIS, J. F. and K. A. REINERT (1997): Applied Methods of Trade Policy Analysis. An Overview. In: FRANCOIS, J. F. and K. A. REINERT (eds.): Applied Methods for Trade Policy Analysis. Cambridge: Cambridge University Press, 3-24.

GALANOPOULOS, K., LINDBERG, E., NILSSON, F. O. L. and Y. SURRY (2007): Agricultural Situation Synthesis Report. Deliverable prepared for the Project SSPE-CT-2004-502459 (STREP) „MEDFROL Market and Trade Policies for the Mediterranean Agriculture: the case of fruit/vegetable and olive oil", December. Available from: http://medfrol.maich.gr, accessed at 02/2007.

GARCIA ÁLVAREZ-COQUE, J. M. (2006a): The Multilateral Trade Negotiations and their Implications for Mediterranean Countries. In: Centre International de Hautes Etudes Agronomiques Méditerranéennes (CIHEAM) (ed.): Agri.Med Agriculture, Fisheries, Food and Sustainable Rural Development in the Mediterranean Region. Annual Report 2006. Paris: CIHEAM, 1-30.

—— (2006b): La Reforma de la OCM de Frutas y Hortalizas: Estudio. Versión Provisional. IP/B/AGRI/ST/2006_161. Dirección General de Políticas Internas de la Unión, Departamento Temático Políticas Estructurales y de Cohesión, Agricultura y Desarrollo Rural, Parliamento Europeo, Brussels.

—— (2002): Agricultural Trade and the Barcelona Process: Is full Liberalisation possible? European Review of Agricultural Economics 20 (3): 399-422.

GARCIA ÁLVAREZ-COQUE, J. M. and M. L. MARTÍ SELVA (2006). A Gravity Approach to Assess the Effects of Association Agreements on Euromediterranean Trade of Fruits and Vegetables. Working paper No. 06/15, prepared for the Project (STREP) „Agricultural TRADE Agreements".

GARCIA ÁLVAREZ-COQUE, J. M., MARTÍNEZ-GÓMEZ, V. and M. VILLANUEVA (2007): F&V Trade Model to assess Euro-Med Agreements: An Application to the Fresh Tomato Market. Contributed Paper presented at the 1st Mediterranean Conference of Agro-Food Social Scientists and 103rd Seminar of the European Association of Agricultural Economists (EAAE), Barcelona, April 23 – 25.

—— (2006): Modelling Euro-Mediterranean Agricultural Trade. Working paper No. 2006/05, prepared for the Project (STREP) „Agricultural TRADE Agreements".

GAVIN, B. (2005): The Euro-Mediterranean Partnership: an Experiment in North-South-South Integration. Intereconomics 40 (6): 353-360.

GRAMS, M. (2004): Analyse der EU-Milchmarktpolitik bei Unsicherheit. Dissertation, Landwirtschaftlich-Gärtnerische Fakultät der Humboldt Universität zu Berlin, Berlin: electronic publication.

GRETHE, H. (2005): Auswirkungen einer Integration der Agrarmärkte der Türkei und der EU – Eine Partielle Gleichgewichtsanalyse. In: HAGEDORN, K., NAGEL, U. J. and M. ODENING (eds.): Umwelt- und Produktqualität im Agrarbereich. Schriften der Gesellschaft für Wirtschafts- und Sozialwissenschaften des Landbaues e.V., Bd. 40, Münster-Hiltrup: Landwirtschaftsverlag, 479-488.

—— (2004): Marktintegration ohne Gemeinsame Agrarpolitik: Auswirkungen einer Einbeziehung von Agrarprodukten in die Zollunion mit der EU auf den türkischen Agrarsektor. Agrarwirtschaft 53 (7): 300-313.

—— (2003): Effects of Including Agricultural Products in the Customs Union between Turkey and the EU. Dissertation, Frankfurt am Main: Peter Lang Verlag.

GRETHE, H., NOLTE, S. and S. TANGERMANN (2006): Evolution, Current State and Future of the EU Trade Preferences for Agricultural Products from North-African Countries and Near-East Countries. In: MIJKOVIC, D. (ed.): New Topics in International Agricultural Trade and Development. New York: Nova Science Publishers Inc., 109-133.

—— (2005a): Entwicklung und Zukunft der EU-Agrarhandelspräferenzen für die südlichen und östlichen Mittelmeeranrainerstaaten. Agrarwirtschaft 54 (7): 300-313.

—— (2005b): The Development and Future of EU Agricultural Trade Preferences for North-African and Near-East Countries. Contributed Paper at the XIth Conference of the European Association of Agricultural Economists (EAAE), Copenhagen, August 27-27.

—— (2005c): Evolution, Current State and Future of the EU Trade Preferences for Agricultural Products from North-African Countries and Near-East Countries. Journal of International Agricultural Trade and Development 1 (2): 109-133.

GRETHE, H. and S. TANGERMANN (1998a): The New Euro-Mediterranean Agreements: An Analysis of Trade Preferences in Agriculture. Discussions Paper 9902, Institute of Agricultural Economics, University of Göttingen. Available from: http://www.gwdg.de/ ~uaao/disbeit/d9902.pdf, accessed at 17/07/2003.

—— (1998b): The Euro-Mediterranean Agreements: An Analysis of Trade Preferences in Agriculture. Paper prepared for the Commodities and Trade Division FAO Economic and Social Department, Institute of Agricultural Economics, University of Göttingen.

HANDOUSSA, H. and J.-L. REIFFERS (2003): Femise Report on the Euro-Mediterranean Partnership 2003: Analysis and Proposal of the Euro-Mediterranean Forum of Economic Institutes. Report prepared for the Femise Network, Marseilles.

HARRISON, G. W., RUTHERFORD, T. F. and D. G. TARR (1997): Economic implications for Turkey of a Customs Union with the European Union. European Economic Review 41 (3-5): 861-870.

HARRISON, G. W., RUTHERFORD, T. F. and D. G. TARR (1996): Economic Implications for Turkey of a CU with the European Union. Policy Research Working Paper No. 1599, The World Bank, Washington D.C.

HECKELEI, T. (2002): Calibration and Estimation of Programming Models for Agricultural Supply Analysis. Habilitationsschrift an der Landwirtschaftlichen Fakultät der Rheinischen Friedrich-Wilhelms-Universtität Bonn.

HENRICHSMEYER, W. and H. P. WITZKE (1994): Agrarpolitik Band 2: Bewertung und Willensbildung. Stuttgart: Verlag Eugen Ulmer.

HERTEL, T. W. (ed.) (1998): Global Trade Analysis: Modelling and Applications. Cambridge: Cambridge University Press.

HOEKMAN, B., KONAN, D. and K. MASKUS (2001): Overlapping Free Trade Agreements in the Middle East and North Africa: Economic Incentives and Effects in Egypt. In: DESSUS, S., DEVLIN, J. and R. SAFADI (eds.): Towards Arab and Euro-Med Regional Integration. Paris: OECD, 169-188.

HOSOE, N. (2001): A General Equilibrium Analysis of Jordan's Trade Liberalisation. Journal of Policy Modeling 23 (6): 595-600.

IARC (2005): Sustainability Impacts of the Euro-Mediterranean Free Trade Area. Final Report on the Phase 2 of the Sustainability Impact Assessment of the Euro-Mediterranean Free Trade Area (SIA-EMFTA) Project, Impact Assessment Research Centre, Institute for Development Policy and Management, University of Manchester.

——— (2004): Sustainability Impact Assessment Study of the Euro-Mediterranean Free Trade Area: SIA Methodology. Final report on Phase 1 of the SIA-EMFTA Project prepared for the European Commission under Contract No. EuropeAid/114340/C/SV/CME, Impact Assessment Research Centre, Institute for Development Policy and Management, University of Manchester.

JUST, R. E., HUETH, D. L. and A. SCHMITZ (2004): The Welfare Economics of Public Policy: a Practical Approach to Project and Policy Evaluation. Cheltenham and Northampton: Edward Elgar Publishing.

KALAITZIS, P., DIJK, G. VAN and G. BAOURAKIS (2007): Euro-Mediterranean Supply Chain Developments and Trends in Trade Structures in the Fresh Fruit and Vegetable Sector. Contributed Paper presented at the 1st Mediterranean Conference of Agro-Food Social Scientists and 103rd Seminar of the European Association of Agricultural Economists (EAAE), Barcelona, April 23 – 25.

KARAGIANNIS, G. (2004): The EU Cotton Regime and the Implications of the Proposed Changes for Producer Welfare. Commodity and Trade Policy Research Working Paper No. 9, Commodities and Trade Division, Food and Agriculture Organisation of the United Nations (FAO), Rome.

KAVALLARI A., BORRESCH R. and P. M. SCHMITZ (2005a): Modelling Agricultural Policy Reforms in the Mediterranean Basin: Adjustments of AGRISIM. Contributed paper at the 98th Seminar of the European Association of Agricultural Economists (EAAE), Chania, June 29 – July 2.

—— (2005b): Modelling CAP Reform for the Mediterranean Countries-The Case of Tobacco, Olive Oil and Cotton. In ARFINI F. (ed.): Modelling Agricultural Policies: State of Art and New Challenges. Proceedings of the 89th European Seminar of the European Association of Agricultural Economists. Parma: Monte Università Parma, 694-716.

—— (2005c): CAP Reform and the Mediterranean Member States. Poster Paper presented at the XIth Conference of the European Association of Agricultural Economists (EAAE), Copenhagen, August 27-27.

KEE, H. L., NICITA, A. and M. OLARREAGA (2004): Estimating Trade Restrictiveness Indices. Policy Research Working Paper No. 3840, The World Bank, Washington D.C.

KIRSCHKE, D. and K. JECHLITSCHKA (2002): Angewandte Mikroökonomie und Wirtschaftspolitik mit Excel. München: Verlag Franz Vahlen.

KIRSCHKE, D. and P. M. SCHMITZ (1990): Grundlagen der angewandten Wohlfahrtsökonomie. Wirtschaftswissenschaftliches Studium (WiSt) (7): 328-333.

KLARE K. and H. DOLL (2004): Reform der Gemeinsamen Agrarpolitik unter besonderer Berücksichtigung der Auswirkungen auf die Pachtpreise, Stellungnahmen im Auftrag des Bundesministeriums für Verbraucherschutz, Ernährung und Landwirtschaft. Institut für ländliche Räume, FAL, Braunschweig. Available from: http://www.bw.fal.de/ download/gap-reform_2004.pdf, accessed at 03/06/04.

KONAN, D. E. and K. E. MASKUS (2000): Joint Trade Liberalisation and Tax Reform in a Small Open Economy: the Case of Egypt. Journal of Development Economics 61 (2): 365-392.

KONAN, D. E. and K. E. MASKUS (1997): A Computable General Equilibrium Analysis of Egyptian Trade Liberalisation Scenarios. In: GALAL, A. and B. HOEKMAN (eds.): Regional Partners in Global Markets: Limits and Possibilities of the Euro-Med Agreements. London: Centre for Economic Policy Reasearch (CEPR), Cairo: The Egyptian Center for Economic Studies (ECES), 156-177.

KRUEGER, A. O. (1997): Free Trade Agreements versus Customs Unions. Journal of Development Economics 54 (1): 169-187.

KUHN, A. (2003): From World Market to Trade Flow Modelling - the Re-Designed WATSIM Model. Final Report on the Project: WATSIM AMPS - Applying and

Maintaining the Policy Simulation Version of the World Agricultural Trade Simulation Model, Institute of Agricultural Policy, Market Research and Economic Sociology, University of Bonn, Bonn.

KUIPER, M. (2006): An Economy-wide Perspective on Euro-Mediterranean Trade Agreements with a Focus on Morocco and Tunisia. Conference Paper, Thematic Network on Trade Agreements and European Agriculture, European Network of Agricultural and Rural Policy Research Institutes (ENARPRI), Brussels.

—— (2004): Fifty Ways to leave your Protection: Comparing Applied Models of Euro-Mediterranean Association Agreements. Working paper No. 6, Thematic Network on Trade Agreements and European Agriculture, European Network of Agricultural and Rural Policy Research Institutes (ENARPRI), Brussels.

KURZWEIL M., LEDEBUR, O. VON and P. SALAMON (2003): Review of Trade Agreements and Issues. Working Paper No. 3, Thematic Network on Trade Agreements and European Agriculture, European Network of Agricultural and Rural Policy Research Institutes (ENARPRI), Brussels.

LEDEBUR, O. VON, SALAMON, P. and G. WEBER (2005): Who Is Telling the Truth? Synthetic Uniformity Structured or Econometric Country Specific Models: A Model Comparison Based on the Luxembourg Agreement. In ARFINI F. (ed.): Modelling Agricultural Policies: State of Art and New Challenges, Proceedings of the 89th European Seminar of the European Association of Agricultural Economists (EAAE). Parma: Monte Università Parma, 208-221.

LIAPIS, P. S. (2007): Preferential Trade Agreements: How much Do They Benefit Developing Economies. Paris: OECD.

LIPS, M. (2004): The CAP Mid Term Review and the WTO Doha Round: Analyses for the Netherlands, EU and Accession Countries. Report 6.04.03., Agricultural Economics Research Institute (LEI), The Hague.

LÖFGREN, H., HARRIS, R. L. and S. ROBINSON (2002): A Standard Computable General Equilibrium (CGE) Model in GAMS. Microcomputers in Policy Research No. 5, International Food Policy and Research Institute (IFPRI), Washington D.C.

LÖFGREN, H., EL-SAID, M. and S. ROBINSON (2001): Trade Liberalisation and the Poor: A Dynamic Rural-Urban General Equilibrium Analysis of Morocco. In: DESSUS, S., DEVLIN, J. and R. SAFADI (eds.): Towards Arab and Euro-Med Regional Integration. Paris: OECD, 129-146.

LOPEZ, J. A. (2001): Decoupling: a Conceptual Overview. Mimeo. Organisation for Economic Cooperation and Development (OECD), Paris.

LORCA, A. and J. VICENS (2000): L´Impact de la Libéralisation Commerciale Euro-méditerranéenne dans les Échanges Agricoles. Report for the Femise Network, Universidad Autónoma de Madrid.

LUCKE, B. (2001): Fiscal Impact of Trade Liberalisation: the Case of Jordan and Syria. Report for the Femise Network, Marseilles.

Market Access Database (various years): Applied Tariffs Database. Directorate-General for Trade, European Commission. Available from: http://mkaccdb.eu.int/mkaccdb2/index Publi.htm, accessed at 10/2003-10/2005.

MARTÍNEZ GÓMEZ, V. D. (2007): Assessing Euro-Med Trade Preferences: The Case of Entry Price Reduction. Contributed paper at the 1st Mediterranean Conference of Agro-Food Social Scientists and 103rd Seminar of the European Association of Agricultural Economists, Barcelona, April 23–25.

MARTÍNEZ GÓMEZ, V. D. and J. M. GARCIA ÁLVAREZ-COQUE (2005): Vegetable Trade Flows between the European Union and its Mediterranean Partners: an Analysis of the Influence of Preferences and Competitiveness. New Medit 4 (2): 4-10.

MASALA, C. (2000): Die Euro-Mediterrane Partnerschaft: Geschichte-Struktur-Prozeß. Discussion Paper C68, Center for European Integration Studies, Rheinische Friedrich-Wilhelms-Universität Bonn.

M'BAREK, R. (2002): Der nordafrikanische Agrarsektor im Spannungsfeld einer euromediterranen Freihandelszone: Wohlfahrtsökonomische Auswirkungen einer Liberalisierung des Agrarsektors zwischen der Europäischen Union und den Maghreb-staaten Marokko und Tunesien. Dissertation. Berlin: Logos Verlag.

MEIJL H. VAN and F. VAN TONGEREN (2002): The Agenda 2000 CAP Reform, World Prices and GATT-WTO Export Constraints. European Review of Agricultural Economics 29 (4): 445-470.

MERCENIER, J. and E. YELDAN (1997): On Turkey's Trade Policy: Is a Customs Union with Europe enough? European Economic Review 41 (3-5): 871-880.

MINOT, N., CHEMINGUI, M., THOMAS, M., DEWINA, R. and D. ORDEN (2007): Impact of Trade Liberalisation on Agriculture in the Near East and North Africa. Near East and North Africa Division, International Food Policy Research Institute (IFPRI) and International Fund for Agricultural Development (IFAD), Washington D.C. and Rome.

MUAZ, S., JABARIN, A. S., ASSAF, L., SAHAWNEH, M., EL-ROUDAN, O., AL-SAEED, A., SHAHEEN, H., AL-HINDI, A. and A. SALEH (2003): The Impact of Euro-Mediterranean Partnership on the Agricultural Sectors of Jordan, Palestine, Syria, Lebanon and Egypt (The Case of Horticultural Exports to the EU Markets). Report No. FEM21-03 prepared for the Femise Network, Marseilles.

MURO, P. DE and L. SALVATICI (2001): The Common Agricultural Policy in Multisectoral Models. Working paper No. 11, Osservatorio sulle Politiche Agricole dell' UE, Italian National Institure of Agricultural Economics (INEA), Rome.

NICHOLSON, W. (1995): Microeconomic Theory: Basic Principles and Extensions, 6th edition. Fort Worth: The Dryden Press.

NIELSEN, C. P. (2003): Regional and Preferential Trade Agreements: a Literature Review and Identification of Future Steps. Report No. 155, Fødevareøkonomisk Institut, Copenhagen.

NILSSON, F. O. L., LINDBERG, E. and Y. SURRY (2007): Are the Mediterranean Countries Competitive in Fresh Fruit and Vegetable Exports? Food Economics-Acta Agriculturae Scandinavica C 4 (4): 203-216.

OECD (2006): Producer and Consumer Support Estimates: OECD Database 1986-2005. Agricultural Electronic Data Products, Directorate for Food, Agriculture and Fisheries, Organisation for the Economic Cooperation and Development (OECD), Paris.

OECD (various years): Agricultural Outlook. Directorate for Food, Agriculture and Fisheries, Organisation for the Economic Cooperation and Development (OECD), Paris.

Official Journal of the European Union. L 161, 30/04/2004, 48-96.

Official Journal of the European Union. L 270, 21/10/2003, 1-69.

Official Journal of the European Communities. L 311, 12/12/2000, 1-8.

Official Journal of the European Communities. L 148, 01/06/2001, 3-8.

Official Journal of the European Communities. L 358, 31/12/1998, 17-42.

Official Journal of the European Communities. L 210, 28/07/1998, 32-37.

Official Journal of the European Communities. L 189, 30/07/1996, 1-9.

Official Journal of the European Communities. L 215, 30/07/1992, 70-76.

OLARREAGA, M. and Ç. ÖZDEN (2005): AGOA and Apparel: Who Captures the Tariff Rent in the Presence of Preferential Market Access? The World Economy 28 (1): 63-77.

PANAGARIYA, A. (2000): Preferential Trade Liberalization: The Traditional Theory and New Developments. Journal of Economic Literature 38 (2): 287-331.

PÖYHÖNEN, P. (1963): A Tentative Model for the Volume of Trade between Countries. Weltwirtschaftliches Archiv 90 (1): 93-99.

PRADA, L. DE and J. DEKA (2004): Euro-Med Association Agreements : Implementation Guide. DG External Relations, EU Commission. Available from: http://ec.europa.eu/external_relations/euromed/asso_agree_guide_en.pdf, accessed at 21/08/2008.

PUSTOVIT, N. (2003): EU-Osterweiterung und WTO-Liberalisierung aus Sicht der ukrainischen Agrarwirtschaft – Wirkungsanalyse und Bewertung mit Hilfe eines partiellen Gleichgewichtsmodells. Agrarökonomische Monographien und Sammelwerke, Kiel: Wissenschaftsverlag Vauk Kiel KG.

QUEFELEC, S. (2004): Euro-Mediterranean Trade in Agricultural Products. Statistics in Focus, External Trade, Theme 6-1/2004, Eurostat, European Communities, Brussels.

RADWAN, S. and J.-L. REIFFERS (2005): The Euro-Mediterranean Partnership: 10 years after Barcelona: Achievements and Perspectives. Report prepared for the Femise Network, Marseilles.

—— (2003): The Impact of Agricultural Liberalization in the Context of the Euro-Mediterranean Partnership. Report prepared for the Femise Network, Marseilles.

RAVALLION, M. and M. LOKSHIN (2004): Gainers and Losers from Trade Reform in Morocco. Middle East and North Africa Working Paper Series No. 37, The World Bank, Washington D.C.

REED, M. R. (2001): International Trade in Agricultural Products. New Jersey: Prentice-Hall Inc.

ROBINSON, S. and K. THIERFELDER (2002): Trade Liberalisation and Regional Integration: the Search for large Numbers. The Australian Journal of Agricultural and Resource Economics 46 (4): 585-604.

RONINGEN, V. O. (1997): Multi-Market, Multi-Region Partial Equilibrium Modelling. In: FRANCOIS, J. F. and A. REINERT (eds.): Applied Methods for Trade Policy Analysis. A Handbook. Cambridge: Cambridge University Press, 231-257.

RONINGEN, V. O., SULLIVAN J. and P. DIXIT (1991): Documentation of the Static World Policy Simulation (SWOPSIM) Modelling Framework. Staff Report Nr. AGES 9151, Economic Research Service, US Department of Agriculture, Washington D. C.

RUTHERFORD, T. F., RUTSTROM, E. E. and D. TARR (1997): Morocco's Free Trade Agreement with the EU: A Quantitative Assessment. Economic Modelling 14 (2): 237-269.

—— (1993): Morocco's Free Trade Agreement with the European Community: A Quantitative Assessment. Policy Research Working paper No. 1173, The World Bank, Washington D. C.

SADOULET, E. and A. DE JANVRY (1995): Quantitative Development Policy Analysis. Baltimore: The John Hopkins University Press.

SCKOKAI, P. (2001): The Common Agricultural Policy in Econometric Models. Working Paper No. 10, Osservatorio sulle Politiche Agricole dell' UE, Italian National Institure of Agricultural Economics (INEA), Rome.

SCHMITZ, K. (2002): Simulationsmodell für die Weltagrarmärkte – Modellbeschreibung. In: Schmitz P. M. (ed.): Nutzen – Kosten – Analyse Pflanzenschutz. Kiel: Wissenschaftsverlag Vauk Kiel KG, Anhang III, 117-137.

SEALE, J. Jr., REGMI, A. and J. BERNSTEIN (2003): International Evidence on Food Consumption Patterns. Technical Bulletin No. 1904, Economic Research Service, United States Department of Agriculture, Washington D.C.

SONMEZ, Y., MCDONALD, S. and J. PERRATON (2006): Trade Implications of Turkey's Accession to the EU. Paper prepared for the ECOMOD International Conference on Policy Modelling, Hong Kong, June 28-30.

STOFOROS, C. and G. MERGOS (2004): Modelling EU Tobacco Policy Reform. Applied Economics 36 (19): 2167-2175.

List of references

SULLIVAN, J., RONINGEN, V., LEETMAA, S. and D. GRAY (1992): A 1989 Global Database for the Static World Policy Simulation (SWOPSIM) Modelling Framework. Staff Report No. AGES 9215, Agriculture and Trade Analysis Division, Economic Research Service, U.S. Department of Agriculture, Washington D.C.

TARIC (various years): Online Customs Tariff Database. Directorate-General for Taxation and Customs Union, European Commission. Available from http://ec.europa.eu/ taxation_customs/dds/en/ tarhome.htm, accessed at 10/2003-10/2005.

TINBERGEN, J. (1962): Shaping the World Economy. New York: The Twentieth Century Fund Inc.

TONGEREN, F. VAN, MEIJL, H. VAN and Y. SURRY (2001): Global Models applied to Agricultural and Trade Policies: a Review and Assessment. Agricultural Economics 26 (2): 149-172.

TONGEREN, F. VAN, MEIJL, H. VAN, VEENENDAAL, P., FRANDSEN, S. E., NIELSEN, C. P., STÆHR, M. H. J., BROCKMEIER, M., MANEGOLD, D., FRANCOIS, J., RAMBOUT, M., SURRY, Y., VAITTINEN, R., KERKELA, L., RATINGER, T., THOMSON, K., FRAHAN, B. H. DE, EL MEKKI, A. A. and L. SALVATICI (2000): Review of Agricultural Trade Models: An Assessment of Models with EU Policy Relevance. In: FRANDSEN, S. E. and M. H. J. STÆHR (eds.): Assessment of the GTAP Modelling Framework for Policy Analysis from a European Perspective. Rapport No. 116, Ministeriet for Fødevarer, Landbrug og Fiskeri, Statens Jordbrugs- og Fiskeriøkonomiske Institut, Copenhagen.

TONGEREN, F. VAN and H. VAN MEIJL (1999): Review of Applied Models of International Trade in Agriculture and Related Resource and Environmental Modelling. FAIR6 CT 98-4148 Interim Report No. 1, Report 5.99.11, Agricultural Economics Research Institute (LEI), The Hague.

TRAINS (various years): Trade Analysis and Information System Database. United Nations Conference on Trade and Development (UNCTAD). Accessed through the World Integrated Trade Solution (WITS) at: 11/2005-11/2006.

TSAKOK, I. (1990): Agricultural Price Policy – A practitioner's Guide to Partial-Equilibrium Analysis. New York: Cornell University Press.

UMA (2008): Objectifs et Missions. L'Union du Maghreb arabe, Website information. Available from: http://www.maghrebarabe.org/fr/obj.cfm, accessed at 22/01/2008.

UNCTAD (2004): User Manual and Handbook on the Agricultural Trade Policy Simulation Model (ATPSM). United Nations Conference on Trade and Development.

WAHL, O., WEBER, G. and K. FROHBERG (2000): Documentation of the Central and Eastern European Countries Agricultural Simulation Model (CEEC-ASIM Version 1.0). Discussion Paper No. 27, Institute of Agricultural Development in Central and Eastern Europe (IAMO), Halle.

WESTHOFF, P. and R. YOUNG (2001): The Status of FAPRI's EU Modelling Effort. In: HECKELEI, T., WITZKE, H. P. and W. HENRICHSMEYER (eds.): Agricultural Sector Modelling and Policy Information Systems. Kiel: Wissenschaftsverlag Vauk Kiel KG, 256-263.

WONNACOTT, P. and R. WONNACOTT (2005): What's the Point of Reciprocal Trade Negotiations? Exports, Imports, and Gains from Trade. World Economy 28 (1): 1-20.

WTO (2007): Regional Trade Agreements. World Trade Organisation (WTO), Website information. Available from: http://www.wto.org/english/tratop_e/region_e/region_e. htm, accessed at 12/10/2007.

WTO notifications (2008): Regional Trade Agreements notified to the GATT/WTO and in Force as of 2 October 2007. World Trade Organisation (WTO). Available from: http://www.wto.org/english/tratop_e/region_e/regfac_e.htm, accessed at 22/01/2008.

—— (2006): Pan-Arab Free Trade Area Agreement: Notification from Saudi Arabia. WT/REG223/N/1, Committee on Regional Trade Agreements, World Trade Organisation (WTO), Geneva.

YAMAZAKI, F. (1996): Potential Erosion of Trade Preferences in Agricultural Products. Food Policy 21 (4/5): 409-417.

Annex

A. To Chapter two

Calculation of the Value of Preference Margins

The estimation of preference margins has been the subject of a number of empirical studies. ALEXANDRAKI and LANKES (2004) estimated the preference margins for middle-income developing countries, while earlier YAMAZAKI (1996) estimated the value of preferences granted by the EU, the USA and Japan to all their preferential trade partners based on 1992 data. GRETHE and TANGERMANN (1998a) calculated the VPM for all agricultural commodities among Mediterranean countries and more recently MARTÍNEZ GÓMEZ (2007) did the same for Moroccan clementines.

The VPM is defined as the difference between the preferential and the world market price:

$$VPM = (P_P - P_W) q_P \qquad (A.1)$$

with P_P price perceived by preferential exporters

P_W world market price

q_P quantity exported by the preferential country (or quantity imported to the preference-granting country)

A graphic illustration follows in Figure A.1. ID_A is the import demand curve for a certain commodity of country A. ES_B is the export supply curve of this commodity from country B and ES_C is the export supply curve for the same commodity from country C. Country C can also be considered as the rest of world. In order to simplify the figure it is assumed that country B has an upward-sloping export supply curve, whereas country C has a perfectly elastic export supply curve. At the beginning a MFN tariff (t_{MFN}) is applied to the exports from both B and C to country A, meaning that country A imports quantity Q_1 from country B and quantity (Q_3-Q_1) from country C, with a price P_W+t_{MFN}. It should be noted that P_W is the world market price. Under this regime the export supply curves of countries B and C are formed in relation to the new domestic price in country A, i.e. to ES_B+t_{MFN} and to ES_C+t_{MFN} respectively. Assuming that country A only gives preferential access to country B, then a new preferential tariff, t_P, is applied to imports from B, resulting in an increase of exports from B country to country A from Q_1 to Q_2 at the expense of country C, whose exports to country A are reduced to (Q_3-Q_2). Due to the preferential regime the export supply curve of country B shifts, as shown in the figure by the curve ES_B+T_P. The resultant preference margin under this new preferential regime equals the area a+b, the area estimated by equation A.1. It is equal to the tariff revenue no longer collected by country

A under the new preferential regime, which it grants instead to exporters from country B.

Figure A.1: Illustration of preference margin effects

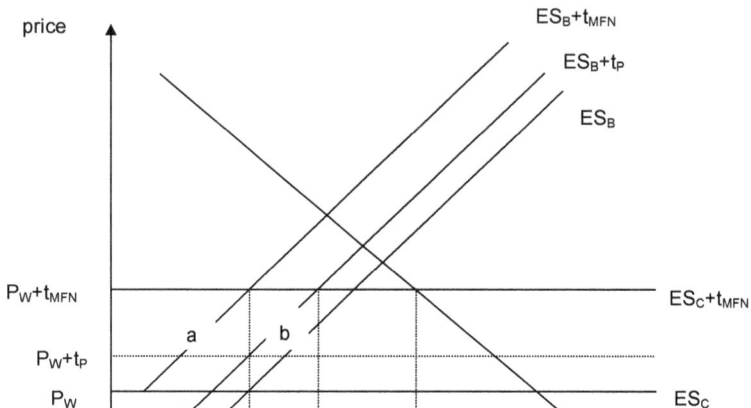

Source: own compilation based on LIAPIS, 2007, pp. 147-148; FRANCOIS et al., 2006; REED, 2001, pp. 97-103.

When working with trade statistics and with a large number of commodities and countries, it is often easier to utilise the value of trade flows instead of individual price and quantity data, and this approach is adopted within most research (ALEXANDRAKI and LANKES, 2004; YAMAZAKI, 1996; GRETHE and TANGERMANN, 1998a; MARTÍNEZ GÓMEZ, 2007; LIAPIS, 2007). LIAPIS (2007, p. 148) asks whether or not the value of trade flows is influenced by the value of the preference margin. To avoid this caveat (and following the studies mentioned above) the calculations of the value of trade flow are adjusted by the MFN tariff rate. Following GRETHE and TANGERMANN (1998a) it is assumed that both preferential and non-preferential commodities (i.e. commodities to which MFN tariff rates are applied) are homogenous and are thus sold on EU markets at the same domestic price. This is described in the following equation:

$P_W(1+t_{MFN}) = P_P(1+t_P)$ (A.2)

with t_{MFN} MFN ad valorem import tariff to the EU

    t_P preferential ad valorem import tariff to the EU

Annex

By combining equations A.1 and A.2, equation A.3 is derived:

$$VPM = \frac{(t_{MFN} - t_P)}{1 + t_{MFN}} P_p q_p \qquad (A.3)$$

While $P_p q_p$ have been replaced by the value of the trade flow.

This equation was applied when calculating the VPM granted by the EU to commodities imported from the MPCs, but only those commodities included in the AGRISIM model. The tables below provide the detailed results. Due to data availability the value of trade flow used for these calculations is the import value, derived from the TRAINS database through the WITS application.

Table A.1: Value of Preference Margins resulting from the Euro-Mediterranean Association Agreements in '000 US$ (1999)

Commodity (HS 1996)		Country Morocco	Turkey	rest of MPCs	Algeria	Egypt	Israel	of which Jordan	Lebanon	Libya	Syria	Tunisia
0201	Bovine meat	n.a[2]	0.00	0.00	n.a	n.a	0.00	n.a	n.a	n.a	n.a	n.a
0203	Pig meat	n.a	n.a	0.00	n.a	n.a	n.a	n.a	n.a	n.a	n.a	n.a
0207	Poultry meat	8.03	0.00	522.16	n.a	n.a	522.16	n.a	n.a	n.a	0.00	n.a
0401	Milk and cream, not concentrated	0.00	0.00	0.00	n.a	n.a	0.00	n.a	n.a	n.a	n.a	n.a
0402	Milk and cream, concentrated	n.a	0.00	0.00	n.a	n.a	0.00	n.a	0.00	0.00	n.a	0.00
0702	Tomatoes	0.00	0.00	0.00	0.00	0.00	0.00	0.00	n.a	0.00	0.00	0.00
080510	Oranges	33168.58	0.00	16250.72	n.a	1135.16	10017.52	n.a	0.00	n.a	0.00	5098.04
080810	Apples	380.96	0.00	1.69	0.11	0.27	0.00	n.a	n.a	n.a	0.92	0.38
1001	Wheat and meslin	n.a	0.00	0.00	n.a	0.00	0.00	n.a	n.a	n.a	n.a	n.a
1003	Barley	n.a	n.a	0.00	n.a	0.00	0.00	n.a	n.a	n.a	n.a	n.a
1005	Maize (corn)	0.00	0.00	0.00	n.a	0.00	0.00	n.a	n.a	n.a	n.a	0.00
1006	Rice	n.a	0.00	0.00	n.a	0.00	0.00	0.00	0.00	0.00	n.a	n.a
1007	Grain sorghum	n.a	n.a	0.00	n.a	0.00	0.00	n.a	n.a	n.a	n.a	0.00
1008	Other cereals	0.00	0.00	0.00	n.a	0.00	n.a	n.a	n.a	n.a	n.a	0.00
1201	Soya beans	n.a	0.00	0.00	n.a	0.00	0.00	n.a	n.a	n.a	n.a	n.a
1204	Linseed	n.a	0.00	0.00	n.a	0.00	0.00	n.a	n.a	n.a	n.a	n.a
1206	Sunflower seeds	0.00	0.00	0.00	n.a	0.00	0.00	n.a	n.a	n.a	n.a	n.a
1207	Other oil seeds	0.00	0.00	0.00	n.a	0.00	0.00	0.00	0.00	n.a	0.00	n.a
1507	Soya-bean oil	65.85	n.a	0.93	n.a	n.a	n.a	n.a	0.93	n.a	n.a	n.a
1509	Olive oil	0.00	0.00	0.00	n.a	0.00	0.00	n.a	0.00	n.a	0.00	0.00
1512	Sunflower, safflower or cotton-seed oil	n.a	0.00	0.00	n.a	n.a	0.00	n.a	n.a	n.a	n.a	n.a
2401	Unprocessed tobacco	0.00	0.00	0.00	0.00	n.a	0.00	n.a	0.00	n.a	0.00	0.00
5201	Cotton-not carded or combed	0.00	0.00	0.00	n.a	0.00	0.00	n.a	n.a	n.a	0.00	0.00
170111	Cane sugar	n.a	n.a	0.00	n.a	n.a	0.00	n.a	n.a	n.a	n.a	n.a
	Sum	33623.42	0.00	16775.50	0.11	1135.43	10539.68	0.00	0.93	0.00	0.92	5098.42

Notes: for the period 1998-2003 import duties (into the EU) where reported only for 1999 and 2003 and for Libya only for 1999; n.a= non-available import duty for this commodity; MPCs: Mediterranean Partner Countries

Source: Own calculations based on reported import duties derived from TRAINS and bilateral trade flows derived from COMTRADE

Annex

Table A.2: Value of Preference Margins resulting from the Euro-Mediterranean Association Agreements in '000 US$ (2003)

Commodity (HS 1996)		Morocco	Turkey	rest of MPCs	Algeria	Egypt	Israel	of which Jordan	Lebanon	Libya	Syria	Tunisia
0201	Bovine meat	n.a[2]	n.a	0.00	0.00	n.a	0.00	n.a	n.a	n.a	0.00	n.a
0203	Pig meat	0.00	n.a	0.00	n.a	n.a	0.00	n.a	0.00	n.a	n.a	n.a
0207	Poultry meat	n.a	111.79	0.00	n.a	n.a	0.00	n.a	n.a	n.a	n.a	n.a
0401	Milk and cream, not concentrated	n.a	0.00	0.00	0.00	n.a	n.a	n.a	n.a	n.a	n.a	n.a
0402	Milk and cream, concentrated	n.a	0.00	0.00	0.00	n.a	0.00	n.a	n.a	n.a	n.a	n.a
0702	Tomatoes	58370.40	19387.20	2447.66	1.36	444.09	0.00	316.72	3.11	n.a	0.00	1682.38
080510	Oranges	48831.55	5780.60	30046.55	n.a	4725.44	18548.43	n.a	36.15	n.a	n.a	6736.53
080810	Apples	-0.03	343.39	-0.03	n.a	-0.02	0.00	n.a	n.a	n.a	-0.02	n.a
1001	Wheat and meslin	n.a	133.26	0.87	n.a	n.a	0.00	n.a	0.87	n.a	0.00	n.a
1003	Barley	n.a	0.00	0.00	n.a	n.a	n.a	n.a	n.a	n.a	n.a	n.a
1005	Maize (corn)	0.00	0.00	0.00	n.a	n.a	0.00	n.a	n.a	n.a	n.a	n.a
1006	Rice	n.a	687.73	0.00	n.a	0.00	n.a	n.a	n.a	n.a	0.00	0.00
1007	Grain sorghum	n.a	n.a	0.00	n.a	0.00	n.a	n.a	n.a	n.a	n.a	n.a
1008	Other cereals	n.a	n.a	0.00	n.a	0.00	n.a	n.a	0.00	n.a	0.00	n.a
1201	Soya beans	n.a	n.a	0.00	n.a	n.a	0.00	n.a	n.a	n.a	n.a	n.a
1204	Linseed	0.00	0.00	0.00	n.a	0.00	0.00	0.00	0.00	n.a	n.a	n.a
1206	Sunflower seeds	n.a	0.00	0.00	n.a	0.00	0.00	n.a	n.a	n.a	0.00	n.a
1207	Other oil seeds	0.00	0.00	0.00	n.a	0.00	0.00	n.a	0.00	n.a	0.00	n.a
1507	Soya-bean oil	28.51	2.34	0.00	n.a	n.a	n.a	n.a	n.a	n.a	n.a	n.a
1509	Olive oil	0.00	0.00	0.00	n.a	0.00	0.00	n.a	0.00	n.a	0.00	0.00
1512	Sunflower, safflower or cotton-seed oil	33.98	3.29	8.29	n.a	8.29	n.a	n.a	n.a	n.a	n.a	n.a
2401	Unprocessed tobacco	n.a	0.00	0.00	n.a	0.00	n.a	n.a	0.00	n.a	0.00	n.a
5201	Cotton-not carded or combed	0.00	0.00	0.00	n.a	0.00	0.00	n.a	n.a	n.a	0.00	n.a
170111	Cane sugar	n.a	0.00	0.00	n.a	n.a	n.a	n.a	n.a	n.a	n.a	n.a
	Sum	107264.41	26449.60	32503.34	1.36	5177.80	18548.43	316.72	40.13	0.00	-0.02	8418.91

Notes: for the period 1998-2003 import duties (into the EU) where reported only for 1999 and 2003 and for Libya only for 1999; n.a= non-available import duty for this commodity

Source: Own calculations based on reported import duties derived from TRAINS and bilateral trade flows derived from COMTRADE

Table A.3: Applied tariffs by the MPCs (in %)

Reporter	Partner	Tariff year	HS 1996	Description	Weighted average of AVE
Algeria	Egypt	2001	1207	Other oil seeds	5
Algeria	Egypt	2001	1006	Rice	10.32
Algeria	Jordan	2001	0402	Milk and cream, concentrated	5
Algeria	Syria	2001	5201	Cotton, not carded or combed	5
Algeria	Syria	2001	1207	Other oil seeds	5
Algeria	Syria	2001	1001	Wheat and meslin	2.5
Algeria	Turkey	2001	1003	Barley	7.5
Algeria	Turkey	2001	1207	Other oil seeds	5
Algeria	Turkey	2001	1001	Wheat and meslin	2.5
Egypt	Israel	1998	080810	Apples	40
Egypt	Israel	1998	1206	Sunflower seeds	1
Egypt	Jordan	1998	080810	Apples	40
Egypt	Jordan	1998	070200	Tomatoes	20
Egypt	Lebanon	2002	080810	Apples	40
Egypt	Lebanon	1998	1005	Maize (corn)	1
Egypt	Lebanon	2002	1509	Olive oil	12.5
Egypt	Syria	2002	080810	Apples	40
Egypt	Syria	2002	5201	Cotton, not carded or combed	5
Egypt	Syria	2002	100110	Durum wheat	1
Egypt	Syria	2002	080510	Oranges	40
Egypt	Tunisia	1998	1509	Olive oil	12.5
Egypt	Turkey	2002	080810	Apples	40
Egypt	Turkey	2002	1005	Maize (corn)	1
Egypt	Turkey	1998	1509	Olive oil	12.5
Egypt	Turkey	1998	1507	Soya-bean oil	8.67
Egypt	Turkey	2002	1206	Sunflower seeds	1
Israel	Egypt	2005	1006	Rice	0
Israel	Jordan	2005	1509	Olive oil	0
Israel	Jordan	2005	070200	Tomatoes	0
Israel	Turkey	2005	5201	Cotton, not carded or combed	0
Israel	Turkey	2005	1005	Maize (corn)	0
Israel	Turkey	2005	1509	Olive oil	8
Jordan	Israel	2001	0402	Milk and cream, concentrated	5
Jordan	Lebanon	2001	080510	Oranges	32.5
Jordan	Lebanon	2001	1507	Soya-bean oil	17.5
Jordan	Syria	2001	5201	Cotton, not carded or combed	0

Table A.3: -continued-

Reporter	Partner	Tariff year	HS 1996	Description	Weighted average of AVE
Jordan	Syria	2001	080510	Oranges	32.5
Jordan	Syria	2001	1207	Other oil seeds	10
Jordan	Syria	2001	1512	Sunflower-seed, safflower or cotton-seed oil	30
Jordan	Syria	2001	2401	Unmanufactured tobacco	20
Jordan	Turkey	2001	1003	Barley	0
Jordan	Turkey	2001	1005	Maize (corn)	5
Jordan	Turkey	2001	080510	Oranges	32.5
Jordan	Turkey	2001	1207	Other oil seeds	5
Jordan	Turkey	2001	1205	Rape or colza seeds	10
Jordan	Turkey	2001	1512	Sunflower-seed, safflower or cotton-seed oil	13.9
Jordan	Turkey	2001	1001	Wheat and meslin	0
Lebanon	Egypt	2001	0207	Meat and edible offal, of the poultry	5
Lebanon	Egypt	2001	1207	Other oil seeds	5
Lebanon	Egypt	2001	1006	Rice	5
Lebanon	Egypt	2001	1512	Sunflower-seed, safflower or cotton-seed oil	15
Lebanon	Egypt	2001	0702	Tomatoes	70
Lebanon	Jordan	2001	0207	Meat and edible offal, of the poultry	5
Lebanon	Jordan	2001	0702	Tomatoes	70
Lebanon	Syria	2001	1008	Other cereals	5
Lebanon	Syria	2001	5201	Cotton, not carded or combed	0
Lebanon	Syria	2001	1207	Other oil seeds	0.49
Lebanon	Syria	2001	1512	Sunflower-seed, safflower or cotton-seed oil	15
Lebanon	Syria	2001	0702	Tomatoes	70
Lebanon	Turkey	2001	080810	Apples	70
Lebanon	Turkey	2001	1008	Other cereals	5
Lebanon	Turkey	2001	5201	Cotton, not carded or combed	0
Lebanon	Turkey	2001	1005	Maize (corn)	0
Lebanon	Turkey	2001	0402	Milk and cream, concentrated	5
Lebanon	Turkey	2001	1207	Other oil seeds	4.22
Lebanon	Turkey	2001	1507	Soya-bean oil	15
Lebanon	Turkey	2001	1206	Sunflower seeds	0
Lebanon	Turkey	2001	1512	Sunflower-seed, safflower or cotton-seed oil	15
Lebanon	Turkey	2001	1001	Wheat and meslin	0

Table A.3: -continued-

Reporter	Partner	Tariff year	Commodity HS 1996	Description	Weighted average of AVE
Libya	Tunisia	2002	080810	Apples	30
Libya	Tunisia	2002	1003	Barley	0
Libya	Tunisia	2002	1007	Grain sorghum	0
Libya	Tunisia	2002	1005	Maize (corn)	0
Libya	Tunisia	2002	0207	Meat and edible offal, of the poultry	50
Libya	Tunisia	2002	1507	Soya-bean oil	15
Libya	Tunisia	2002	1206	Sunflower seeds	12.5
Libya	Turkey	2002	1003	Barley	0
Morocco	Egypt	2001	5201	Cotton, not carded or combed	2.5
Morocco	Egypt	2001	0207	Meat and edible offal, of the poultry	131.5
Morocco	Egypt	2001	1207	Other oil seeds	28.5
Morocco	Egypt	2001	1006	Rice	98.11
Morocco	Lebanon	2001	1005	Maize (corn)	17.5
Morocco	Lebanon	2001	1207	Other oil seeds	22.12
Morocco	Lebanon	2001	1206	Sunflower seeds	24.8
Morocco	Libya	2001	1007	Grain sorghum	13.75
Morocco	Syria	2001	5201	Cotton, not carded or combed	2.5
Morocco	Tunisia	2001	0207	Meat and edible offal, of the poultry	117.2
Morocco	Tunisia	2001	1509	Olive oil	54.5
Morocco	Turkey	2001	1003	Barley	24.88
Morocco	Turkey	2001	5201	Cotton, not carded or combed	2.5
Morocco	Turkey	2001	1509	Olive oil	54.5
Morocco	Turkey	2001	1207	Other oil seeds	28.5
Morocco	Turkey	2001	2401	Unmanufactured tobacco	17.5
Morocco	Turkey	2001	1001	Wheat and meslin	29.09
Syria	Egypt	2002	1006	Rice	7
Syria	Egypt	2002	1507	Soya-bean oil	4
Syria	Egypt	2002	0702	Tomatoes	15
Syria	Jordan	2002	0702	Tomatoes	15
Syria	Lebanon	2002	080510	Oranges	15
Syria	Lebanon	2002	1507	Soya-bean oil	4.79
Syria	Turkey	2002	1003	Barley	1
Tunisia	Egypt	2004	1007	Grain sorghum	34.33
Tunisia	Egypt	2003	1204	Linseed	43
Tunisia	Egypt	2002	1005	Maize (corn)	0
Tunisia	Egypt	2004	1006	Rice	27

Annex 201

Table A.3: -continued-

Reporter	Partner	Tariff year	HS 1996	Description	Weighted average of AVE
Tunisia	Egypt	2003	1206	Sunflower seeds	75
Tunisia	Jordan	2002	1509	Olive oil	156
Tunisia	Lebanon	2004	100890	Other cereals	100
Tunisia	Libya	2002	1007	Grain sorghum	34.33
Tunisia	Libya	2004	080510	Oranges	200
Tunisia	Libya	1998	1206	Sunflower seeds	43
Tunisia	Libya	2002	0702	Tomatoes	170
Tunisia	Morocco	2004	0207	Meat and edible offal, of the poultry	73
Tunisia	Morocco	1998	0203	Meat of swine	43
Tunisia	Morocco	1998	1206	Sunflower seeds	43
Tunisia	Syria	2004	5201	Cotton, not carded or combed.	0
Tunisia	Syria	2003	100110	Durum wheat	79
Tunisia	Turkey	2002	1003	Barley	84
Tunisia	Turkey	2004	5201	Cotton, not carded or combed	0
Tunisia	Turkey	1998	100110	Durum wheat	17
Tunisia	Turkey	2004	1005	Maize (corn)	0
Tunisia	Turkey	2003	2401	Unmanufactured tobacco	25
Turkey	Egypt	2003	5201	Cotton, not carded or combed	0
Turkey	Egypt	2003	1006	Rice	44.83
Turkey	Egypt	2003	1206	Sunflower seeds	8
Turkey	Israel	2003	5201	Cotton, not carded or combed	0
Turkey	Israel	1999	100110	Durum wheat	25
Turkey	Israel	2003	1005	Maize (corn)	0
Turkey	Israel	2003	1507	Soya-bean oil	25
Turkey	Israel	1999	1206	Sunflower seeds	19
Turkey	Jordan	2003	080810	Apples	60.9
Turkey	Jordan	2003	0207	Meat and edible offal, of the poultry	62.23
Turkey	Jordan	2003	100820	Millet	20
Turkey	Jordan	2003	080510	Oranges	54.6
Turkey	Jordan	1999	1006	Rice	35
Turkey	Lebanon	1999	1006	Rice	35
Turkey	Syria	2003	5201	Cotton, not carded or combed	0
Turkey	Syria	2003	100110	Durum wheat	15
Turkey	Syria	1999	1006	Rice	35
Turkey	Tunisia	2003	2401	Unmanufactured tobacco	25

Source: TRAINS

Table A.4: Export subsidies for agricultural commodities reported by the MPCs to the WTO

WTO member	Year	Commodity	Notified expenditures in US$	Quantity notified of subsidised exports in tonnes	Percent of allowed quantity used in %
Israel	1997	Citrus fruits	562000	19900	5
Israel	1997	Flowers fresh	4348000	415[1]	55
Israel	1997	Fruits other than citrus	989000	50	0
Israel	1997	Vegetables, fresh	4600	17600	20
Israel	1998	Flowers fresh	1409000	244[1]	33
Israel	2000	Citrus fruits	561000	50000	13
Israel	2000	Flowers fresh	3181000	354000[2]	49
Israel	2000	Fruits other than citrus	599000	35000	70
Israel	2000	Goose liver	83000	60	21
Israel	2000	Vegetables, fresh	459000	40000	48
Israel	2001	Citrus fruits	380000	60000	16
Israel	2001	Cotton	886000	600	1
Israel	2001	Flowers fresh	2320000	526000000[3]	75
Israel	2001	Fruits other than citrus	1488000	36000	73
Israel	2001	Goose liver	25000	100	44
Israel	2001	Vegetables, fresh	836000	78200	96
Morocco	1997	Flowers	277598.92	1571	n.a
Morocco	1997	Fruit	14020.147	86	n.a
Morocco	1997	Vegetables	451033.337	1286	n.a
Morocco	1998	Flowers	260965.142	1108	n.a
Morocco	1998	Fruit	45849.471	112	n.a
Morocco	1998	Vegetables	223594.682	729	n.a
Morocco	1999	Flowers	84805.299	496	n.a
Morocco	1999	Fruit	22280.209	50	n.a
Morocco	1999	Vegetables	290746.688	1165	n.a
Morocco	2000	Flowers	91189.386	512	n.a
Morocco	2000	Fruit	68111.744	196	n.a
Morocco	2000	Vegetables	42044.287	137	n.a
Tunisia	1997	Citrus fruit	210151.38	3169	n.a
Tunisia	1997	Dates	349065.004	4505	n.a
Tunisia	1997	Sardines	3807658.059	3000000[4]	n.a
Tunisia	1997	Tomato double concentrate	4338379.341	13478	n.a
Tunisia	1997	Wine	54318.789	614000[5]	n.a
Tunisia	1998	Citrus fruit	207929.52	3169	n.a

Annex 203

Table A.4: -continued-

WTO member	Year	Commodity	Notified expenditures in US$	Quantity notified of subsidized exports in tonnes	Percent of allowed quantity used in %
Tunisia	1998	Dates	345374.45	4505	n.a
Tunisia	1998	Sardines	376740.09	3000000[4]	n.a
Tunisia	1998	Tomato double concentrate	5085462.56	18239	n.a
Tunisia	1998	Wine	53744.49	614000[5]	n.a
Tunisia	1999	Citrus fruit	269893.355	5536	n.a
Tunisia	1999	Dates	906480.722	13279	n.a
Tunisia	1999	Potatoes	242821.985	3938	n.a
Tunisia	1999	Tomato double concentrate	4019688.269	19322	n.a
Tunisia	1999	Wine	41837.572	506243[5]	n.a
Tunisia	2000	Citrus fruit	220890.165	6667	n.a
Tunisia	2000	Dates	944149.318	14772	n.a
Tunisia	2000	Potatoes	88513.999	4500	n.a
Tunisia	2000	Tomato double concentrate	4712132.089	23266	n.a
Tunisia	2000	Wine	46733.668	651222	n.a
Tunisia	2001	Citrus fruit	442925.496	7117	n.a
Tunisia	2001	Dates	429254.956	19182	n.a
Tunisia	2001	Potatoes	73820.916	1035	n.a
Tunisia	2001	Tomato double concentrate	40328.093	27660	n.a
Tunisia	2001	Wine	2379050	593760	n.a
Turkey	1997	Apples	0	47581	72
Turkey	1997	Citrus fruit (Oranges,Mandarins, Lemons,Grapefruit)	9500	180783	68
Turkey	1997	Creams	817780	136	78
Turkey	1997	Cut flowers (fresh)	352100	87064076[3]	100
Turkey	1997	Eggs	3706600	50300000[3]	97
Turkey	1997	Fruit juices (concentrated)	1314336	21550	100
Turkey	1997	Fruits (frozen)	0	13691	100
Turkey	1997	Homogenised fruit preparations	0	6989	82
Turkey	1997	Meat of the Poultry	0	1569	68
Turkey	1997	Olive oil	2235933	1531	7
Turkey	1997	Onion-Dried	0	106473	80
Turkey	1997	Potatoes	5470	16351	52

Table A.4: -continued-

WTO member	Year	Commodity	Notified expenditures in US$	Quantity notified of subsidized exports in tonnes	Percent of allowed quantity used in %
Turkey	1997	Potatoes (frozen and fried)	1114050	99	1
Turkey	1997	Prepared or preserved fish, crustaceans and molluscs	15435200	7427	91
Turkey	1997	Preserves, pastes	0	280640	100
Turkey	1997	Tomatoes	0	114520	100
Turkey	1997	Vegetables (dehydrated)	1586200	864	79
Turkey	1997	Vegetables, frozen (exc. potatoes)	0	14420	100
Turkey	1998	Apples	0	12324	19
Turkey	1998	Citrus fruit (Oranges, Mandarins, Lemons, Grapefruit)	11550	108861	42
Turkey	1998	Creams	747252	165	97
Turkey	1998	Cut flowers (fresh)	215124	83050000[3]	100
Turkey	1998	Eggs	3411900	30732500[3]	60
Turkey	1998	Fruit juices (concentrated)	730004	20070	100
Turkey	1998	Fruits (frozen)	0	7766	60
Turkey	1998	Homogenised fruit preparations	0	8063	100
Turkey	1998	Meat of the Poultry	0	1784	78
Turkey	1998	Natural honey	1002144	0	0
Turkey	1998	Olive oil	0	10439	47
Turkey	1998	Potatoes (frozen and fried)	604650	927	12
Turkey	1998	Prepared or preserved fish, crustaceans and molluscs	15190500	4031	50
Turkey	1998	Preserves, pastes	0	266500	100
Turkey	1998	Vegetables (dehydrated)	1478740	895	83
Turkey	1998	Vegetables, frozen (exc. potatoes)	0	13820	100
Turkey	1999	Apples	0	3663	6
Turkey	1999	Chocolate and other food preparations containing chocolate	0	21350	100

Table A.4: -continued-

WTO member	Year	Commodity	Notified expenditures in US$	Quantity notified of subsidized exports in tonnes	Percent of allowed quantity used in %
Turkey	1999	Cut flowers (fresh)	141350	79064000[3]	100
Turkey	1999	Eggs	3099600	50482000[3]	100
Turkey	1999	Fruit juices (concentrated)	1035184	18450	99
Turkey	1999	Fruits (frozen)	0	11252	93
Turkey	1999	Homogenised fruit preparations	0	7579	100
Turkey	1999	Macaroni vermicelli	0	9595	31
Turkey	1999	Meat of the Poultry	0	1680	75
Turkey	1999	Olive oil	790636	5102	23
Turkey	1999	Onion-Dried	0	46508	36
Turkey	1999	Potatoes	33495	13821	46
Turkey	1999	Potatoes (frozen and fried)	1666980	609	8
Turkey	1999	Prepared or preserved fish, crustaceans and molluscs	13882440	7938	100
Turkey	1999	Preserves, pastes	0	252408	100
Turkey	1999	Vegetables (dehydrated)	1266594	1060	100
Turkey	1999	Vegetables, frozen (exc. potatoes)	0	11949	90
Turkey	2000	Apples	n.a	2031	3
Turkey	2000	Chocolate and other food preparations containing chocolate	n.a	20420	100
Turkey	2000	Cut flowers (fresh)	160160	58499526[3]	77932
Turkey	2000	Eggs	2665611	22880050[3]	46
Turkey	2000	Fruit juices (concentrated)	654410	15867	93
Turkey	2000	Fruits (frozen)	n.a	7113	62
Turkey	2000	Homogenised fruit preparations	n.a	7050	99
Turkey	2000	Macaroni vermicelli	n.a	10206	37
Turkey	2000	Meat of the Poultry	n.a	1735	78
Turkey	2000	Olive oil	566695	10000	47
Turkey	2000	Onion-Dried	n.a	33335	26
Turkey	2000	Potatoes	n.a	29850	100

Table A.4: -continued-

WTO member	Year	Commodity	Notified expenditures in US$	Quantity notified of subsidized exports in tonnes	Percent of allowed quantity used in %
Turkey	2000	Prepared or preserved fish, crustaceans and molluscs	13103750	7810	100
Turkey	2000	Preserves, pastes	n.a	238250	100
Turkey	2000	Vegetables (dehydrated)	1261400	1045	100
Turkey	2000	Vegetables, frozen (exc. potatoes)	n.a	11900	93

Notes: 1: in millions of units; 2: in thousands of units; 3: in units; 4: in tins; 5: in bottles; n.a: not available

Source: WTO notifications

Annex 207

B. To Chapter five

AGRISIM Simulation Results

Table B.1: Base Run projections of product balances and prices

		Supply (in 1000 t)	Food demand (in 1000 t)	Farm gate price (in US$/t)	Border price (in US$/t)	Producer incentive price (in US$/t)
	WHEA	4609	3643	112	110	129
	COAR	10651	2024	106	99	122
	RICE	816	635	180	180	221
	OILS	349	633	841	841	928
	SUGA	949	1303	266	152	277
	MILK	7082	6910	286	179	324
	BEEF	594	616	1859	768	2210
ESP	PORK	3049	2656	1395	1137	1427
	POUL	1073	1078	957	643	988
	COTT	102	94	1271	1271	2228
	TOBA	42	89	4814	4814	6046
	OLIO	1599	653	1762	1762	2092
	APPL	962	759	583	547	583
	ORAN	2822	1479	550	479	550
	TOMA	3731	1826	634	634	634
	WHEA	2499	1621	112	110	124
	COAR	2310	442	106	99	116
	RICE	138	103	180	180	220
	OILS	9	103	973	973	1028
	SUGA	315	328	266	152	276
	MILK	1920	2143	286	179	334
	BEEF	52	254	1859	768	2048
GRE	PORK	143	390	1395	1137	1427
	POUL	145	213	957	643	988
	COTT	456	167	810	810	1358
	TOBA	136	72	2700	2700	3942
	OLIO	540	334	1162	1162	1673
	APPL	243	190	140	132	140
	ORAN	1022	479	331	289	331
	TOMA	1821	1294	637	637	637
	WHEA	6155	8701	112	110	138
ITA	COAR	10985	1455	106	99	122
	RICE	1183	596	180	180	228

Table B.1: - continued -

		Supply (in 1000 t)	Food demand (in 1000 t)	Farm gate price (in US$/t)	Border price (in US$/t)	Producer incentive price (in US$/t)
ITA	OILS	322	712	852	852	959
	SUGA	1283	1605	266	152	284
	MILK	12317	12965	286	179	317
	BEEF	1025	1378	1859	768	2119
	PORK	1544	2426	1395	1137	1427
	POUL	1140	1065	957	643	988
	COTT	0	275	1460	1460	2185
	TOBA	130	114	1723	1723	2904
	OLIO	588	837	1627	1626	2395
	APPL	2340	1363	507	476	507
	ORAN	1857	1722	478	416	478
	TOMA	6530	3657	1036	1036	1036
E12	WHEA	71738	28175	112	110	127
	COAR	65849	12702	106	99	125
	RICE	232	2312	180	180	228
	OILS	4280	5644	699	699	778
	SUGA	11968	9952	266	152	282
	MILK	104516	89726	286	179	309
	BEEF	5213	5132	1859	768	2298
	PORK	13584	11215	1395	1137	1427
	POUL	6914	5963	957	643	988
	COTT	0	426	895	895	1626
	TOBA	43	617	2576	2576	3841
	OLIO	42	286	1614	1614	2764
	APPL	6129	8673	694	652	694
	ORAN	222	2803	6911	6017	6911
	TOMA	2982	5374	1461	1461	1461
E10	WHEA	22799	8657	111	108	129
	COAR	34328	4402	93	86	113
	RICE	8	348	89	89	134
	OILS	1423	1294	511	511	597
	SUGA	2517	2552	155	88	175
	MILK	22256	12476	302	189	325
	BEEF	817	234	2408	1012	2746
	PORK	3202	3300	1207	983	1249
	POUL	1719	1412	1085	730	1123
	COTT	0	184	1295	1295	1295
	TOBA	36	106	3426	3426	3426

Annex 209

Table B.1: - continued -

		Supply (in 1000 t)	Food demand (in 1000 t)	Farm gate price (in US$/t)	Border price (in US$/t)	Producer incentive price (in US$/t)
E10	OLIO	4	22	2504	2504	2504
	APPL	3643	1752	149	140	149
	ORAN	37	317	421	367	421
	TOMA	604	925	585	585	585
BUR	WHEA	11836	4692	138	104	138
	COAR	13084	1715	130	108	130
	RICE	8	145	5129	4771	5129
	OILS	528	422	615	621	615
	SUGA	74	739	34	15	34
	MILK	6456	5857	205	166	205
	BEEF	217	220	1126	1257	1126
	PORK	708	792	1082	1245	1082
	POUL	392	479	1294	788	1294
	COTT	3	52	1290	1281	1290
	TOBA	51	70	3281	2776	3281
	OLIO	0	5	2198	1966	2198
	APPL	550	553	703	586	703
	ORAN	0	70	715	634	715
	TOMA	1045	1073	1215	972	1215
MOR	WHEA	3332	5459	208	162	208
	COAR	1281	2239	123	123	123
	RICE	37	37	679	321	679
	OILS	9	418	1959	1959	1959
	SUGA	488	954	352	352	352
	MILK	1165	817	1305	607	1305
	BEEF	150	150	1987	1987	1987
	PORK	1	1	802	802	802
	POUL	253	258	1362	651	1362
	COTT	0	38	1080	1053	1080
	TOBA	6	16	2467	2099	2467
	OLIO	68	60	2959	1915	2959
	APPL	228	209	552	358	552
	ORAN	708	408	332	332	332
	TOMA	881	604	401	401	401
TUR	WHEA	19099	13869	130	139	132
	COAR	10333	2557	108	91	111
	RICE	360	694	92	92	92
	OILS	256	601	1027	818	1048

Table B.1: - continued -

		Supply (in 1000 t)	Food demand (in 1000 t)	Farm gate price (in US$/t)	Border price (in US$/t)	Producer incentive price (in US$/t)
TUR	SUGA	2070	1811	293	230	303
	MILK	9374	9155	207	179	208
	BEEF	335	332	2205	718	2228
	PORK	0	0	1310	1310	1310
	POUL	627	622	1014	849	1014
	COTT	900	1324	611	611	611
	TOBA	145	160	1659	1328	1659
	OLIO	117	111	1842	1404	1842
	APPL	2449	2056	356	356	356
	ORAN	1250	1011	325	325	325
	TOMA	8432	5277	257	257	257
MPC	WHEA	14655	23120	166	155	166
	COAR	14750	9555	80	80	80
	RICE	5229	4998	169	169	169
	OILS	69	1129	1813	1813	1813
	SUGA	1430	4900	110	109	110
	MILK	9941	10907	695	551	695
	BEEF	893	1318	2258	2095	2258
	PORK	17	27	878	863	878
	POUL	1709	1708	1596	1331	1596
	COTT	689	527	957	957	957
	TOBA	57	286	2521	2433	2521
	OLIO	430	420	2573	1798	2573
	APPL	1237	1149	500	469	500
	ORAN	3041	2351	326	305	326
	TOMA	9894	8878	486	355	486

Source: Own simulations with AGRISIM

Table B.2: Custom duties and agricultural subsidies in Base Year (in US$ million)

		Custom levies	Direct subsidies	Input subsidies	General subsidies
ESP	WHEA	-8	75	5	0
	COAR	-50	98	7	0
	RICE	-24	41	8	0
	OILS	0	316	20	0
	SUGA	-12	2	6	0
	MILK	-9	4	8	0

Table B.2: - continued -

		Custom levies	Direct subsidies	Input subsidies	General subsidies
ESP	BEEF	-120	846	156	0
	PORK	121	8	28	0
	POUL	13	6	28	0
	COTT	0	1864	0	0
	TOBA	0	2449	0	0
	OLIO	0	579	0	0
	APPL	0	0	0	0
	ORAN	68	0	0	0
	TOMA	0	0	0	0
GRE	WHEA	1	75	5	0
	COAR	-9	98	7	0
	RICE	-3	41	8	0
	OILS	0	316	20	0
	SUGA	2	2	6	0
	MILK	-32	4	8	0
	BEEF	-182	846	156	0
	PORK	-73	8	28	0
	POUL	-20	6	28	0
	COTT	0	1068	0	0
	TOBA	0	2470	0	0
	OLIO	0	975	0	0
	APPL	0	0	0	0
	ORAN	17	0	0	0
	TOMA	0	0	0	0
ITA	WHEA	-12	75	5	0
	COAR	-31	98	7	0
	RICE	-51	41	8	0
	OILS	0	316	20	0
	SUGA	-17	2	6	0
	MILK	-186	4	8	0
	BEEF	-553	846	156	0
	PORK	-213	8	28	0
	POUL	24	6	28	0
	COTT	0	1327	0	0
	TOBA	0	2325	0	0
	OLIO	0	1298	0	0
	APPL	23	0	0	0
	ORAN	4	0	0	0
	TOMA	0	0	0	0

Table B.2: - continued -

		Custom levies	Direct subsidies	Input subsidies	General subsidies
E12	WHEA	10	75	5	0
	COAR	-105	98	7	0
	RICE	138	41	8	0
	OILS	0	316	20	0
	SUGA	441	2	6	0
	MILK	895	4	8	0
	BEEF	-768	846	156	0
	PORK	766	8	28	0
	POUL	368	6	28	0
	COTT	0	1327	0	0
	TOBA	0	2479	0	0
	OLIO	0	819	0	0
	APPL	-129	0	0	0
	ORAN	-1551	0	0	0
	TOMA	0	0	0	0
E10	WHEA	13	7	5	0
	COAR	39	3	4	0
	RICE	0	0	0	0
	OILS	3	20	29	0
	SUGA	16	6	8	0
	MILK	1781	5	5	1
	BEEF	1493	38	29	10
	PORK	-69	14	38	2
	POUL	121	17	35	1
	COTT	0	0	0	0
	TOBA	0	0	0	0
	OLIO	64	0	0	0
	APPL	13	0	0	0
	ORAN	-20	0	0	0
	TOMA	11	0	0	0
BUR	WHEA	13	0	0	0
	COAR	-1	0	0	0
	RICE	-52	0	0	0
	OILS	-1	0	0	0
	SUGA	-14	0	0	0
	MILK	-1	0	0	0
	BEEF	2	0	0	0
	PORK	14	0	0	0
	POUL	-41	0	0	0
	COTT	0	0	0	0

Table B.2: - continued -

		Custom levies	Direct subsidies	Input subsidies	General subsidies
BUR	TOBA	-7	0	0	0
	OLIO	-1	0	0	0
	APPL	-2	0	0	0
	ORAN	-6	0	0	0
	TOMA	-32	0	0	0
MOR	WHEA	-149	0	0	0
	COAR	0	0	0	0
	RICE	-1	0	0	0
	OILS	0	0	0	0
	SUGA	0	0	0	0
	MILK	-6	0	0	0
	BEEF	0	0	0	0
	PORK	0	0	0	0
	POUL	-2	0	0	0
	COTT	-1	0	0	0
	TOBA	-3	0	0	0
	OLIO	2	0	0	0
	APPL	2	0	0	0
	ORAN	0	0	0	0
	TOMA	0	0	0	0
TUR	WHEA	-10	0	2	0
	COAR	-7	0	3	0
	RICE	0	0	0	0
	OILS	-71	0	22	0
	SUGA	57	0	9	0
	MILK	0	1	0	0
	BEEF	1	0	23	0
	PORK	0	0	0	0
	POUL	1	0	0	0
	COTT	0	0	0	0
	TOBA	18	0	0	0
	OLIO	47	0	0	0
	APPL	0	0	0	0
	ORAN	0	0	0	0
	TOMA	0	0	0	0
MPC	WHEA	-147	0	0	0
	COAR	0	0	0	0
	RICE	0	0	0	0
	OILS	0	0	0	0
	SUGA	-7	0	0	0

Table B.2: - continued -

		Custom levies	Direct subsidies	Input subsidies	General subsidies
MPC	MILK	-194	0	0	0
	BEEF	-67	0	0	0
	PORK	0	0	0	0
	POUL	0	0	0	0
	COTT	0	0	0	0
	TOBA	-15	0	0	0
	OLIO	81	0	0	0
	APPL	-1	0	0	0
	ORAN	9	0	0	0
	TOMA	12	0	0	0

Source: Own calculations with AGRISIM

Table B.3: Changes in commodity balances (deviation from BA in %)

		Supply					Food demand				
		SC1	SC2	SC3	SC4	SC5	SC1	SC2	SC3	SC4	SC5
ESP	WHEA	1	1	2	2	3	0	0	0	2	3
	COAR	1	1	1	1	1	0	0	0	5	10
	RICE	0	0	0	6	1	0	0	0	-9	-2
	OILS	1	1	1	2	3	0	0	1	0	0
	SUGA	0	0	0	0	0	0	0	0	8	20
	MILK	0	0	0	0	0	0	0	0	2	4
	BEEF	1	1	1	-6	-19	0	0	0	10	40
	PORK	0	0	-1	-5	-9	0	0	0	4	7
	POUL	0	0	-1	-11	-21	0	0	0	14	29
	COTT	-4	-4	-4	-4	-4	0	0	0	0	-1
	TOBA	-1	-1	-1	-1	0	0	0	0	-1	-2
	OLIO	-1	-1	0	-1	0	-1	-1	-3	-2	-4
	APPL	0	0	0	-1	-2	0	0	0	0	1
	ORAN	0	0	0	-3	-6	0	0	0	1	3
	TOMA	0	0	3	2	3	0	0	0	0	-1
GRE	WHEA	-2	-2	-2	1	3	0	0	0	1	2
	COAR	-2	-2	-2	-2	-3	0	0	0	2	5
	RICE	1	1	1	6	2	0	0	0	-9	-3
	OILS	0	0	-1	2	4	0	0	1	1	1
	SUGA	0	0	0	0	0	0	0	0	8	20
	MILK	0	0	0	0	0	0	0	0	2	5
	BEEF	2	2	2	-5	-20	0	0	0	11	43
	PORK	-1	-1	-2	-5	-9	0	0	0	1	-2
	POUL	-1	0	-1	-12	-22	0	0	0	13	28

Annex 215

Table B.3: - continued -

		Supply					Food demand				
		SC1	SC2	SC3	SC4	SC5	SC1	SC2	SC3	SC4	SC5
GRE	COTT	-4	-4	-4	-4	-3	0	0	0	-1	-2
	TOBA	-4	-4	-4	-4	-3	0	0	0	-2	-4
	OLIO	-2	-2	-2	-2	-2	-1	-1	-5	-3	-5
	APPL	0	0	0	4	8	0	0	0	-1	-3
	ORAN	0	0	-1	-2	-4	0	0	0	1	1
	TOMA	0	0	3	2	3	0	0	0	0	-1
ITA	WHEA	2	2	2	2	2	0	0	0	2	4
	COAR	1	1	1	-1	-3	0	0	0	5	11
	RICE	0	0	0	6	1	0	0	0	-10	-3
	OILS	1	1	1	2	2	0	0	0	0	0
	SUGA	0	0	0	0	0	0	0	0	8	20
	MILK	0	0	0	0	0	0	0	0	2	5
	BEEF	0	0	0	-6	-21	0	0	0	11	43
	PORK	0	0	0	-4	-6	0	0	0	2	0
	POUL	0	0	0	-11	-22	0	0	0	13	27
	COTT	-3	-3	-3	-3	-3	0	0	0	0	-1
	TOBA	-7	-7	-7	-6	-5	0	0	-1	-3	-5
	OLIO	-1	-1	-1	-1	-1	-1	-1	-4	-2	-4
	APPL	0	0	0	-1	-1	0	0	0	0	0
	ORAN	0	0	-1	-3	-6	0	0	0	1	2
	TOMA	0	0	2	1	2	0	0	0	0	0
E12	WHEA	0	0	0	0	0	0	0	0	1	2
	COAR	0	0	0	-1	-2	0	0	0	2	4
	RICE	0	0	0	6	1	0	0	0	-9	-2
	OILS	0	0	0	0	0	0	0	0	0	0
	SUGA	0	0	0	0	0	0	0	0	8	20
	MILK	0	0	0	0	0	0	0	0	2	4
	BEEF	0	0	0	-6	-19	0	0	0	11	44
	PORK	0	0	0	-4	-6	0	0	0	2	1
	POUL	0	0	0	-11	-21	0	0	0	14	28
	COTT	-4	-4	-4	-4	-4	0	0	0	-1	-1
	TOBA	-5	-5	-4	-4	-3	0	0	0	-2	-4
	OLIO	-1	-1	-1	-1	-1	-1	-1	-4	-3	-4
	APPL	0	0	0	-1	-2	0	0	0	0	1
	ORAN	0	0	0	-5	-10	0	0	0	1	3
	TOMA	0	0	1	1	1	0	0	0	0	0

Table B.3: - continued -

		Supply					Food demand				
		SC1	SC2	SC3	SC4	SC5	SC1	SC2	SC3	SC4	SC5
E10	WHEA	0	0	0	0	0	0	0	0	2	3
	COAR	0	0	0	0	0	0	0	0	3	8
	RICE	0	0	0	0	1	0	0	0	-1	-2
	OILS	0	0	0	0	1	0	0	0	0	-1
	SUGA	0	0	0	-2	-5	0	0	0	5	13
	MILK	0	0	0	0	0	0	0	2	18	44
	BEEF	0	0	0	-5	-11	0	1	3	35	115
	PORK	0	0	0	-1	-2	0	0	0	7	13
	POUL	0	0	0	-4	-9	0	0	0	20	41
	COTT	0	0	0	0	0	0	0	0	0	0
	TOBA	0	0	0	-1	-2	0	0	0	-1	-2
	OLIO	0	0	1	-1	-2	0	0	-1	-1	-1
	APPL	0	0	0	3	7	0	0	0	-1	-3
	ORAN	0	0	-1	-3	-5	0	0	0	1	3
	TOMA	0	0	3	2	3	0	0	-1	0	-1
BUR	WHEA	0	-4	-4	-4	-4	0	-1	-1	1	3
	COAR	0	3	3	3	3	0	-5	-5	-3	-1
	RICE	0	-1	-1	-1	-1	0	1	1	1	1
	OILS	0	1	1	1	1	0	0	0	0	-1
	SUGA	0	0	0	0	0	0	16	19	-4	-14
	MILK	0	0	0	0	0	0	-24	-23	-11	7
	BEEF	0	25	25	19	10	0	-72	-72	-62	-39
	PORK	0	8	8	7	6	0	-28	-29	-23	-19
	POUL	0	-4	-4	-9	-13	0	23	23	48	76
	COTT	0	33	32	28	21	0	0	0	0	0
	TOBA	0	-1	-1	-1	-1	0	5	5	4	3
	OLIO	0	-1	-1	-1	-1	0	3	2	2	2
	APPL	0	-9	-9	-10	-11	0	5	5	6	6
	ORAN	0	1	1	-3	-6	0	-1	-1	1	3
	TOMA	0	-32	-31	-31	-31	0	10	9	10	9
MOR	WHEA	0	0	-6	-2	-4	0	0	4	2	5
	COAR	0	0	-11	-2	-7	0	0	-2	0	0
	RICE	0	0	-6	-4	-10	0	0	3	2	5
	OILS	0	0	-1	0	0	0	0	0	0	0
	SUGA	0	0	6	0	1	0	0	-5	0	-1
	MILK	0	0	-6	-6	-14	0	0	0	0	0
	BEEF	0	0	23	2	5	0	0	-24	-1	-4
	PORK	0	0	6	1	3	0	0	-4	-1	-2
	POUL	0	0	-23	-18	-38	0	0	0	0	0

Annex 217

Table B.3: - continued -

		Supply					Food demand				
		SC1	SC2	SC3	SC4	SC5	SC1	SC2	SC3	SC4	SC5
MOR	COTT	0	0	7	5	11	0	0	1	0	0
	TOBA	0	0	-1	0	-1	0	0	2	0	1
	OLIO	0	0	-11	-5	-10	0	0	6	2	6
	APPL	0	0	-26	-12	-26	0	0	18	8	19
	ORAN	0	0	11	3	7	0	0	-6	-2	-4
	TOMA	0	0	4	3	5	0	0	-1	-1	-1
TUR	WHEA	0	0	2	1	1	0	0	-2	0	-1
	COAR	0	0	-12	-3	-6	0	0	3	3	7
	RICE	0	0	5	0	1	0	0	-2	0	-1
	OILS	0	0	-3	-1	-3	0	0	2	1	2
	SUGA	0	0	3	-1	-1	0	0	-3	1	1
	MILK	0	0	25	0	2	0	0	-5	0	0
	BEEF	0	0	-3	-5	-14	0	0	6	8	24
	PORK	0	0	0	0	0	0	0	0	0	0
	POUL	0	0	12	-2	-3	0	0	5	-3	-5
	COTT	0	0	-1	1	2	0	0	0	0	-1
	TOBA	0	0	-2	0	-1	0	0	2	0	1
	OLIO	1	1	-5	-2	-5	0	0	3	1	3
	APPL	0	0	5	2	5	0	0	-3	-1	-3
	ORAN	0	0	11	3	7	0	0	-5	-2	-3
	TOMA	0	0	7	4	8	0	0	-2	-1	-2
MPC	WHEA	0	0	-2	-1	-3	0	0	4	1	2
	COAR	0	0	-2	1	3	0	0	-3	0	0
	RICE	0	0	6	0	1	0	0	-2	0	0
	OILS	0	0	0	0	0	0	0	-1	0	-1
	SUGA	0	0	5	1	2	0	0	-5	-1	-2
	MILK	0	0	12	-4	-9	0	0	-7	3	6
	BEEF	0	0	20	0	1	0	0	-22	-1	-3
	PORK	0	0	5	1	2	0	0	-4	-1	-1
	POUL	0	0	10	-3	-6	0	0	2	0	0
	COTT	0	0	1	1	2	0	0	0	0	-1
	TOBA	0	0	0	0	0	0	0	0	0	0
	OLIO	1	1	-10	-4	-10	0	0	5	2	5
	APPL	0	0	0	-1	-1	0	0	0	0	1
	ORAN	0	0	5	1	2	0	0	-3	0	-1
	TOMA	0	0	-39	-20	-39	0	0	13	5	13

Source: Own simulations with AGRISIM

Table B.4: Net trade effects (in 1000 t)

		BA	SC1	SC2	SC3	SC4	SC5
	WHEA	-3011	-2952	-2945	-2925	-3032	-3039
	COAR	-4733	-4610	-4656	-4611	-5307	-5840
	RICE	167	168	168	168	271	191
	OILS	-539	-536	-536	-539	-533	-529
	SUGA	-104	-104	-103	-106	-213	-369
	MILK	-71	-71	-74	-85	-207	-370
	BEEF	-26	-18	-17	-17	-121	-388
ESP	PORK	417	410	397	394	165	-56
	POUL	1	-3	1	-5	-266	-534
	COTT	8	4	4	4	5	5
	TOBA	-47	-48	-48	-48	-47	-46
	OLIO	488	479	479	504	492	506
	APPL	-5	-5	-4	-5	-15	-24
	ORAN	957	957	957	943	859	758
	TOMA	1255	1255	1263	1359	1322	1383
	WHEA	130	92	95	95	124	155
	COAR	-1007	-1047	-1053	-1050	-1125	-1198
	RICE	14	15	15	15	33	19
	OILS	-99	-100	-100	-100	-100	-100
	SUGA	16	16	16	16	-11	-51
	MILK	-297	-297	-298	-302	-346	-404
	BEEF	-167	-166	-165	-165	-197	-287
GRE	PORK	-297	-298	-300	-300	-309	-303
	POUL	-68	-69	-68	-69	-114	-159
	COTT	289	272	272	272	274	276
	TOBA	62	57	57	57	59	61
	OLIO	209	201	201	217	209	218
	APPL	44	44	45	44	55	68
	ORAN	405	405	405	397	386	365
	TOMA	344	344	347	395	375	403
	WHEA	-4641	-4531	-4525	-4498	-4746	-4857
	COAR	-2217	-2135	-2186	-2152	-2989	-3721
	RICE	521	524	524	523	650	553
	OILS	-497	-495	-495	-497	-494	-492
ITA	SUGA	-153	-153	-153	-156	-287	-480
	MILK	-1727	-1727	-1734	-1756	-2014	-2357
	BEEF	-353	-349	-348	-348	-566	-1150
	PORK	-882	-885	-894	-895	-985	-991
	POUL	75	73	76	71	-198	-461
	COTT	-275	-275	-275	-275	-274	-272

Table B.4: - continued -

		BA	SC1	SC2	SC3	SC4	SC5
ITA	TOBA	16	7	7	8	11	16
	OLIO	-231	-229	-229	-207	-218	-205
	APPL	740	740	741	740	722	708
	ORAN	70	70	70	58	6	-61
	TOMA	3173	3173	3181	3288	3243	3306
E12	WHEA	9746	9733	9802	9988	9032	9033
	COAR	-746	-744	-888	-766	-2463	-3662
	RICE	-2296	-2296	-2296	-2298	-2051	-2228
	OILS	-2167	-2169	-2169	-2172	-2155	-2139
	SUGA	3866	3866	3871	3848	3032	1833
	MILK	8665	8665	8622	8491	6974	4948
	BEEF	-85	-84	-81	-80	-958	-3300
	PORK	2366	2367	2316	2303	1679	1394
	POUL	974	975	995	963	-595	-2154
	COTT	-426	-425	-425	-425	-423	-420
	TOBA	-266	-266	-266	-265	-257	-244
	OLIO	-268	-265	-265	-257	-261	-256
	APPL	-3037	-3037	-3034	-3037	-3135	-3223
	ORAN	-1736	-1736	-1736	-1737	-1789	-1846
	TOMA	-2843	-2843	-2840	-2803	-2818	-2797
E10	WHEA	1928	1926	1936	1952	1668	1420
	COAR	3741	3741	3727	3671	2914	1821
	RICE	-358	-358	-358	-358	-355	-351
	OILS	153	153	153	154	166	180
	SUGA	214	214	219	203	35	-254
	MILK	8016	8016	7960	7794	5501	1949
	BEEF	538	538	534	526	419	180
	PORK	-98	-97	-112	-111	-353	-586
	POUL	194	194	198	196	-158	-537
	COTT	-183	-183	-183	-183	-183	-183
	TOBA	-44	-44	-44	-44	-43	-43
	OLIO	-18	-18	-18	-17	-18	-17
	APPL	1116	1116	1122	1116	1249	1421
	ORAN	-301	-301	-301	-302	-306	-310
	TOMA	-373	-373	-371	-350	-359	-346
BUR	WHEA	422	421	21	22	-145	-316
	COAR	-7	-6	835	827	625	352
	RICE	-144	-144	-146	-146	-146	-146
	OILS	106	106	110	110	113	116
	SUGA	-710	-710	-833	-854	-676	-604

Table B.4: - continued -

		BA	SC1	SC2	SC3	SC4	SC5
BUR	MILK	-171	-171	1373	1307	563	-587
	BEEF	-10	-10	200	199	166	97
	PORK	-89	-88	192	192	143	98
	POUL	-83	-83	-209	-211	-345	-497
	COTT	-49	-49	-49	-49	-49	-49
	TOBA	-15	-14	-19	-19	-18	-18
	OLIO	-5	-5	-5	-5	-5	-5
	APPL	-13	-13	-91	-91	-101	-107
	ORAN	-74	-74	-73	-73	-75	-76
	TOMA	-131	-131	-537	-522	-528	-519
MOR	WHEA	-3208	-3209	-3208	-3600	-3400	-3589
	COAR	-1770	-1770	-1772	-1797	-1774	-1784
	RICE	-3	-3	-3	-6	-5	-9
	OILS	-411	-411	-411	-411	-411	-411
	SUGA	-466	-466	-466	-386	-461	-453
	MILK	-10	-10	-10	-63	-61	-131
	BEEF	0	0	0	70	5	14
	PORK	0	0	0	0	0	0
	POUL	-4	-4	-4	-63	-51	-100
	COTT	-38	-38	-38	-38	-38	-38
	TOBA	-10	-9	-9	-10	-10	-10
	OLIO	2	2	2	-8	-3	-8
	APPL	8	8	8	-86	-35	-87
	ORAN	259	259	259	356	287	318
	TOMA	207	207	210	250	233	256
TUR	WHEA	1291	1287	1292	1646	1418	1551
	COAR	-169	-171	-164	-2922	-704	-1348
	RICE	-358	-358	-358	-325	-355	-350
	OILS	-343	-344	-344	-362	-352	-361
	SUGA	901	901	902	1022	878	858
	MILK	-155	-155	-181	2511	-135	62
	BEEF	3	2	1	-29	-42	-122
	PORK	0	0	0	0	0	0
	POUL	5	5	6	48	7	15
	COTT	-424	-422	-422	-430	-412	-394
	TOBA	55	55	56	49	54	53
	OLIO	107	108	108	98	104	99
	APPL	222	222	224	393	300	390
	ORAN	143	143	143	323	196	254
	TOMA	642	642	687	948	1007	1315

Table B.4: - continued -

		BA	SC1	SC2	SC3	SC4	SC5
MPC	WHEA	-13941	-13945	-13936	-15495	-14310	-14659
	COAR	-12513	-12517	-12547	-13777	-11852	-11122
	RICE	110	110	111	476	140	184
	OILS	-1125	-1126	-1126	-1119	-1122	-1118
	SUGA	-4414	-4414	-4413	-4093	-4353	-4274
	MILK	-1395	-1395	-1407	475	-2098	-2835
	BEEF	-413	-413	-415	51	-402	-366
	PORK	-10	-10	-10	-8	-9	-9
	POUL	-3	-2	0	134	-59	-100
	COTT	182	183	183	186	191	202
	TOBA	-173	-173	-173	-174	-172	-171
	OLIO	104	107	107	41	77	41
	APPL	-42	-42	-41	-44	-54	-64
	ORAN	431	431	431	641	462	499
	TOMA	99	99	137	-4541	-2167	-4481

Notes: Net trade is calculated as the difference between supply and demand. When supply>demand a region is considered to be a net exporter and when supply<demand a net importer.

Source: Own simulations with AGRISIM

Table B.5: Changes in prices (deviation from BA in %)

		Producer's incentive price					Farm gate price					Border price				
		SC1	SC2	SC3	SC4	SC5	SC1	SC2	SC3	SC4	SC5	SC1	SC2	SC3	SC4	SC5
ESP	WHEA	3	3	3	0	-3	0	0	0	-4	-7	0	0	0	-3	-5
	COAR	2	2	2	-2	-5	0	0	0	-5	-8	0	0	0	-1	-1
	RICE	0	0	0	18	4	0	0	0	22	4	0	0	0	1	4
	OILS	2	2	2	3	3	0	0	0	0	1	0	0	0	0	1
	SUGA	0	0	0	-14	-29	0	0	0	-15	-31	0	0	0	8	21
	MILK	3	2	1	-9	-22	0	0	-2	-14	-28	0	-1	-2	6	16
	BEEF	3	3	3	-10	-33	0	0	0	-15	-42	0	0	-3	13	41
	PORK	0	0	0	-6	-12	0	0	0	-6	-13	0	0	0	3	7
	POUL	0	0	0	-15	-27	0	0	0	-15	-28	0	0	0	2	7
	COTT	-27	-27	-27	-26	-26	0	0	0	1	2	0	0	0	1	2
	TOBA	-20	-20	-20	-19	-17	0	0	0	2	4	0	0	0	2	4
	OLIO	-6	-6	-3	-5	-2	2	2	6	4	6	2	2	6	4	6
	APPL	0	0	0	-1	-2	0	0	0	-1	-2	0	0	0	2	4
	ORAN	0	0	-1	-4	-8	0	0	-1	-4	-8	0	0	-1	3	6
	TOMA	0	0	2	1	2	0	0	2	1	2	0	0	2	1	2
GRE	WHEA	9	9	9	5	3	0	0	0	-4	-7	0	0	0	-3	-5
	COAR	4	4	4	0	-4	0	0	0	-5	-8	0	0	0	-1	-1
	RICE	2	2	2	20	5	0	0	0	22	4	0	0	0	1	4
	OILS	6	6	6	6	7	0	0	0	0	1	0	0	0	0	1
	SUGA	1	1	1	-13	-29	0	0	0	-15	-31	0	0	0	8	21
	MILK	20	19	18	8	-4	0	0	-2	-14	-28	0	0	-2	6	16
	BEEF	3	3	3	-11	-35	0	0	0	-15	-42	0	-1	-3	13	41

Annex 223

Table B.5: - continued -

		Producer incentive price					Farm gate price					Border price				
		SC1	SC2	SC3	SC4	SC5	SC1	SC2	SC3	SC4	SC5	SC1	SC2	SC3	SC4	SC5
GRE	PORK	0	0	0	-6	-12	0	0	0	-6	-13	0	0	0	3	7
	POUL	0	0	0	-15	-27	0	0	0	-15	-28	0	0	0	2	7
	COTT	-24	-24	-24	-23	-22	0	0	0	1	3	0	0	0	1	3
	TOBA	-31	-31	-30	-29	-26	1	1	1	4	8	1	1	1	4	8
	OLIO	-13	-13	-9	-11	-9	3	3	9	6	10	3	3	9	6	10
	APPL	0	0	0	5	10	0	0	0	5	10	0	0	0	8	17
	ORAN	0	0	-1	-2	-5	0	0	-1	-2	-5	0	0	-1	4	9
	TOMA	0	0	2	1	2	0	0	2	1	2	0	0	2	1	2
	WHEA	3	3	3	0	-3	0	0	0	-4	-7	0	0	0	-3	-5
	COAR	1	1	1	-3	-6	0	0	0	-5	-8	0	0	0	-1	-1
	RICE	1	1	1	18	4	0	0	0	22	4	0	0	0	1	4
	OILS	2	2	2	2	3	0	0	0	0	1	0	0	0	0	1
	SUGA	1	1	0	-13	-28	0	0	0	-15	-31	0	0	0	8	21
	MILK	1	1	0	-11	-23	0	0	-2	-14	-28	0	0	-2	6	16
	BEEF	1	1	1	-12	-36	0	0	0	-15	-42	0	-1	-3	13	41
ITA	PORK	0	0	0	-6	-12	0	0	0	-6	-13	0	0	0	3	7
	POUL	0	0	0	-15	-27	0	0	0	-15	-28	0	0	0	2	7
	COTT	-19	-19	-19	-19	-18	0	0	0	1	2	0	0	0	1	2
	TOBA	-39	-39	-39	-37	-33	1	1	1	6	12	1	1	1	6	12
	OLIO	-14	-14	-11	-13	-11	2	2	6	4	7	2	2	6	4	7
	APPL	0	0	0	-1	-2	0	0	0	-1	-2	0	0	0	2	5
	ORAN	0	0	-1	-4	-7	0	0	-1	-4	-7	0	0	-1	3	7
	TOMA	0	0	1	1	1	0	0	1	1	1	0	0	1	1	1

Table B.5: - continued -

		Producer incentive price					Farm gate price					Border price				
		SC1	SC2	SC3	SC4	SC5	SC1	SC2	SC3	SC4	SC5	SC1	SC2	SC3	SC4	SC5
	WHEA	0	0	2	1	2	0	0	0	-4	-7	0	0	0	-3	-5
	COAR	0	0	0	-3	-6	0	0	0	-5	-8	0	0	0	-1	-1
	RICE	0	0	0	-4	-7	0	0	0	22	4	0	0	0	1	4
	OILS	0	0	0	17	3	0	0	0	1	1	0	0	0	1	1
	SUGA	0	0	0	1	1	0	0	0	-15	-31	0	0	0	8	21
	MILK	0	0	0	-14	-29	0	0	-2	-14	-28	0	0	-2	6	16
	BEEF	0	0	-1	-13	-25	0	0	0	-15	-42	0	0	-3	13	41
E12	PORK	0	0	0	-12	-34	0	0	0	-6	-13	0	0	0	3	7
	POUL	0	0	0	-6	-12	0	0	0	-15	-28	0	0	0	2	7
	COTT	0	0	0	-15	-27	0	0	0	1	3	0	0	0	1	3
	TOBA	-26	-26	-26	-26	-25	1	1	1	4	8	1	1	1	4	8
	OLIO	-32	-32	-32	-30	-27	2	2	6	4	7	2	2	6	4	7
	APPL	-8	-8	-5	-6	-5	0	0	0	-1	-3	0	0	0	2	4
	ORAN	0	0	0	-1	-3	0	0	0	-6	-13	0	0	-1	0	0
	TOMA	0	0	0	-6	-13	0	0	1	0	1	0	0	0	0	1
	WHEA	0	0	0	-3	-6	0	0	0	-4	-7	0	0	0	-3	-5
	COAR	0	0	0	-4	-7	0	0	0	-5	-9	0	0	0	-1	-2
	RICE	0	0	0	2	6	0	0	0	3	8	0	0	0	3	8
E10	OILS	0	0	0	1	1	0	0	0	1	2	0	0	0	1	2
	SUGA	0	0	-1	-9	-20	0	0	-1	-10	-22	0	0	-1	14	36
	MILK	0	0	-1	-13	-26	0	0	-1	-14	-28	0	0	-1	6	15
	BEEF	0	-1	-2	-19	-39	-1	-1	-2	-22	-45	-1	-1	-2	10	31
	PORK	0	0	0	-6	-11	0	0	0	-6	-12	0	0	0	4	8

Annex 225

Table B.5: - continued -

		Producer incentive price					Farm gate price					Border price				
		SC1	SC2	SC3	SC4	SC5	SC1	SC2	SC3	SC4	SC5	SC1	SC2	SC3	SC4	SC5
	POUL	0	0	0	-15	-28	0	0	0	-15	-29	0	0	0	1	6
	COTT	0	0	0	1	2	0	0	0	1	2	0	0	0	1	2
	TOBA	0	0	1	3	6	0	0	1	3	6	0	0	1	3	6
E10	OLIO	1	1	4	3	4	1	1	4	3	4	1	1	4	3	4
	APPL	0	0	0	4	9	0	0	0	4	9	0	0	0	8	16
	ORAN	0	0	-1	-4	-6	0	0	-1	-4	-6	0	0	-1	4	7
	TOMA	0	0	2	1	2	0	0	2	1	2	0	0	2	1	2
	WHEA	0	-17	-17	-20	-22	0	-23	-23	-26	-29	0	0	0	-3	-5
	COAR	0	0	0	-4	-7	0	-10	-10	-14	-18	0	0	0	-1	-1
	RICE	0	-6	-6	-6	-6	0	-7	-7	-7	-7	0	0	0	0	0
	OILS	0	12	12	13	14	0	1	1	2	2	0	0	0	1	1
	SUGA	0	42	38	76	101	0	-24	-27	11	35	0	1	-4	86	213
	MILK	0	50	48	33	16	0	29	27	12	-5	0	0	-2	7	17
	BEEF	0	183	180	124	58	0	164	161	105	39	0	-1	-2	8	25
BUR	PORK	0	41	41	32	23	0	41	41	31	23	0	0	0	3	7
	POUL	0	-9	-9	-23	-35	0	-9	-9	-23	-36	0	0	0	1	6
	COTT	0	-1	-1	0	1	0	-1	-1	0	1	0	0	0	1	2
	TOBA	0	-15	-15	-12	-9	0	-15	-15	-12	-9	0	1	1	3	7
	OLIO	2	-9	-6	-8	-6	2	-9	-6	-8	-6	2	2	5	3	6
	APPL	0	-11	-11	-13	-13	0	-11	-11	-13	-13	0	0	0	2	4
	ORAN	0	2	1	-3	-8	0	2	1	-3	-8	0	0	0	2	4
	TOMA	0	-20	-19	-19	-19	0	-20	-19	-19	-19	0	0	1	1	1

Table B.5: - continued -

		Producer incentive price					Farm gate price					Border price				
		SC1	SC2	SC3	SC4	SC5	SC1	SC2	SC3	SC4	SC5	SC1	SC2	SC3	SC4	SC5
MOR	WHEA	0	0	-21	-13	-25	0	0	-21	-13	-25	0	0	0	-2	-3
	COAR	0	0	8	-1	-1	0	0	8	-1	-1	0	0	0	-1	-1
	RICE	0	0	-34	-26	-52	0	0	-34	-26	-52	0	0	0	1	2
	OILS	0	0	0	0	0	0	0	0	0	0	0	0	0	0	0
	SUGA	0	0	75	4	9	0	0	75	4	9	0	0	0	4	9
	MILK	0	0	-26	-25	-51	0	0	-26	-25	-51	0	0	0	2	5
	BEEF	0	0	161	5	16	0	0	161	5	16	0	0	-1	5	16
	PORK	0	0	22	5	10	0	0	22	5	10	0	0	0	5	10
	POUL	0	0	-29	-25	-49	0	0	-29	-25	-49	0	0	0	2	7
	COTT	0	0	-2	0	0	0	0	-2	0	0	0	0	0	1	2
	TOBA	1	1	-14	-3	-7	1	1	-14	-3	-7	1	1	1	5	10
	OLIO	2	2	-32	-15	-32	2	2	-32	-15	-32	2	2	5	3	6
	APPL	0	0	-31	-15	-31	0	0	-31	-15	-31	0	0	0	3	6
	ORAN	0	0	14	4	8	0	0	14	4	8	0	0	-1	4	8
	TOMA	0	0	3	2	3	0	0	3	2	3	0	0	3	2	3
TUR	WHEA	0	0	10	1	3	0	0	10	1	3	0	0	0	-2	-4
	COAR	0	0	-9	-9	-17	0	0	-9	-9	-17	0	0	0	-1	-1
	RICE	0	0	40	3	8	0	0	40	3	8	0	0	0	3	8
	OILS	0	0	-20	-9	-19	0	0	-20	-10	-19	0	0	0	1	1
	SUGA	0	0	36	-6	-10	0	0	37	-6	-11	0	0	0	6	14
	MILK	0	0	36	-1	0	0	0	36	-1	0	0	0	-1	6	16
	BEEF	0	-1	-17	-24	-53	0	-1	-17	-25	-53	0	-1	-3	14	44
	PORK	0	0	22	3	6	0	0	22	3	6	0	0	0	3	6

Table B.5: - continued -

		Producer incentive price					Farm gate price					Border price				
		SC1	SC2	SC3	SC4	SC5	SC1	SC2	SC3	SC4	SC5	SC1	SC2	SC3	SC4	SC5
	POUL	0	0	24	-7	-12	0	0	25	-7	-12	0	0	0	1	5
	COTT	0	0	0	2	4	0	0	0	2	4	0	1	0	2	4
	TOBA	1	1	-19	-4	-8	1	1	-19	-4	-8	1	1	2	7	16
TUR	OLIO	2	2	-18	-8	-18	2	2	-18	-8	-18	2	2	7	5	8
	APPL	0	0	7	3	6	0	0	7	3	6	0	0	0	3	6
	ORAN	0	0	14	4	8	0	0	14	4	8	0	0	-1	4	8
	TOMA	0	0	4	2	5	0	0	4	2	5	0	0	4	2	5
	WHEA	0	0	-4	-5	-9	0	0	-4	-5	-9	0	0	0	-2	-3
	COAR	0	0	8	-1	-2	0	0	8	-1	-2	0	0	0	-1	-2
	RICE	0	0	40	2	4	0	0	40	2	4	0	0	0	2	4
	OILS	0	0	0	0	0	0	0	0	0	0	0	0	0	0	0
	SUGA	0	0	72	11	27	0	0	72	11	27	0	0	0	12	29
	MILK	0	0	26	-9	-17	0	0	26	-9	-17	0	0	0	2	5
	BEEF	0	0	142	1	7	0	0	142	1	7	0	0	-1	5	15
MPC	PORK	0	0	20	3	8	0	0	20	3	8	0	0	0	4	9
	POUL	0	0	24	-8	-14	0	0	24	-8	-14	0	0	0	1	3
	COTT	0	0	0	1	3	0	0	0	1	3	0	0	0	1	3
	TOBA	1	1	-3	2	5	1	1	-3	2	5	1	1	1	4	8
	OLIO	2	2	-26	-12	-26	2	2	-26	-12	-26	2	2	6	4	6
	APPL	0	0	0	-1	-2	0	0	0	-1	-2	0	0	0	2	5
	ORAN	0	0	7	1	2	0	0	7	1	2	0	0	-1	4	9
	TOMA	0	0	-25	-12	-25	0	0	-25	-12	-25	0	0	3	2	3

Source: Own simulations with AGRISIM

Table B.6: Changes of farmer's revenue (% deviations from BA)

		BA	SC1	SC2	SC3	SC4	SC5
ESP	WHEA	0.0	1.4	1.5	2.0	-2.0	-4.4
	COAR	0.0	1.3	1.0	1.3	-3.7	-7.3
	RICE	0.0	0.2	0.2	0.0	29.4	5.5
	OILS	0.0	1.3	1.2	1.2	2.6	4.0
	SUGA	0.0	0.0	0.1	-0.3	-14.8	-31.0
	MILK	0.0	0.0	-0.4	-1.5	-13.7	-27.5
	BEEF	0.0	1.4	1.4	1.4	-19.8	-53.1
	PORK	0.0	-0.2	-0.7	-0.8	-10.9	-20.8
	POUL	0.0	-0.4	-0.1	-0.6	-24.3	-42.7
	COTT	0.0	-4.1	-4.1	-4.1	-3.5	-2.2
	TOBA	0.0	-0.7	-0.6	-0.4	1.3	4.1
	OLIO	0.0	0.7	0.7	5.4	3.1	5.9
	APPL	0.0	0.0	0.1	0.0	-2.1	-3.9
	ORAN	0.0	0.0	0.0	-1.0	-7.0	-13.8
	TOMA	0.0	0.0	0.3	4.4	2.7	5.1
GRE	WHEA	0.0	-1.8	-1.6	-1.4	-2.7	-3.9
	COAR	0.0	-1.9	-2.2	-2.0	-6.9	-11.3
	RICE	0.0	0.6	0.6	0.4	29.8	5.9
	OILS	0.0	-0.5	-0.5	-0.6	2.3	4.9
	SUGA	0.0	0.0	0.1	-0.3	-14.8	-31.0
	MILK	0.0	0.0	-0.4	-1.5	-13.7	-27.5
	BEEF	0.0	1.9	1.9	1.9	-19.8	-53.5
	PORK	0.0	-1.2	-1.7	-1.8	-11.4	-20.7
	POUL	0.0	-0.6	-0.3	-0.8	-25.1	-43.8
	COTT	0.0	-3.5	-3.5	-3.4	-2.4	-0.4
	TOBA	0.0	-3.8	-3.6	-3.3	-0.3	4.5
	OLIO	0.0	0.1	0.1	7.1	3.6	7.8
	APPL	0.0	0.0	0.4	0.0	8.6	19.0
	ORAN	0.0	0.0	0.0	-1.7	-4.1	-8.2
	TOMA	0.0	0.0	0.3	4.4	2.7	5.1
ITA	WHEA	0.0	2.0	2.1	2.6	-2.1	-4.9
	COAR	0.0	0.8	0.5	0.8	-5.6	-10.9
	RICE	0.0	0.3	0.3	0.2	29.3	5.6
	OILS	0.0	1.0	0.9	1.0	2.0	3.1
	SUGA	0.0	0.0	0.1	-0.3	-14.8	-31.0
	MILK	0.0	0.0	-0.4	-1.5	-13.7	-27.5
	BEEF	0.0	0.5	0.5	0.4	-20.7	-53.8
	PORK	0.0	-0.2	-0.7	-0.8	-9.9	-18.2
	POUL	0.0	-0.1	0.2	-0.4	-24.7	-43.4

Annex 229

Table B.6: - continued -

		BA	SC1	SC2	SC3	SC4	SC5
ITA	COTT	0.0	-2.8	-2.8	-2.8	-2.2	-1.1
	TOBA	0.0	-6.4	-6.1	-5.7	-1.1	6.2
	OLIO	0.0	0.6	0.6	5.3	3.0	5.8
	APPL	0.0	0.0	0.1	0.0	-1.6	-2.9
	ORAN	0.0	0.0	0.0	-1.2	-6.3	-12.6
	TOMA	0.0	0.0	0.2	2.7	1.6	3.1
E12	WHEA	0.0	0.0	0.1	0.6	-3.9	-6.8
	COAR	0.0	0.0	-0.3	0.0	-5.5	-9.9
	RICE	0.0	0.0	0.0	-0.1	29.0	5.3
	OILS	0.0	0.0	-0.1	0.0	0.8	1.7
	SUGA	0.0	0.0	0.1	-0.3	-14.8	-31.0
	MILK	0.0	0.0	-0.4	-1.5	-13.7	-27.5
	BEEF	0.0	0.0	0.0	0.0	-20.4	-52.8
	PORK	0.0	0.0	-0.5	-0.6	-9.6	-18.2
	POUL	0.0	0.0	0.3	-0.2	-24.2	-42.9
	COTT	0.0	-4.0	-4.0	-3.9	-3.0	-1.3
	TOBA	0.0	-4.0	-3.9	-3.6	-0.4	4.6
	OLIO	0.0	0.5	0.6	5.5	3.0	6.0
	APPL	0.0	0.0	0.1	0.0	-2.7	-5.0
	ORAN	0.0	0.0	0.0	-0.1	-11.0	-21.4
	TOMA	0.0	0.0	0.1	1.9	1.2	2.2
E10	WHEA	0.0	0.0	0.0	0.4	-3.8	-6.7
	COAR	0.0	0.0	-0.3	0.0	-4.6	-8.8
	RICE	0.0	0.0	0.0	-0.2	3.2	8.9
	OILS	0.0	0.0	-0.1	0.0	1.3	2.8
	SUGA	0.0	0.0	0.3	-0.7	-11.4	-26.3
	MILK	0.0	0.0	-0.4	-1.4	-14.0	-28.0
	BEEF	0.0	-0.1	-0.9	-2.7	-25.1	-51.0
	PORK	0.0	0.0	-0.4	-0.4	-7.2	-13.5
	POUL	0.0	0.0	0.2	-0.1	-19.1	-34.6
	COTT	0.0	0.2	0.2	0.2	0.9	2.2
	TOBA	0.0	0.4	0.5	0.7	1.7	3.8
	OLIO	0.0	1.4	1.4	4.7	1.8	2.2
	APPL	0.0	0.0	0.4	0.0	7.3	17.1
	ORAN	0.0	0.0	0.0	-1.4	-6.3	-11.3
	TOMA	0.0	0.0	0.3	4.8	2.9	5.6
BUR	WHEA	0.0	0.0	-26.3	-26.0	-29.3	-31.7
	COAR	0.0	0.0	-8.0	-7.8	-11.8	-15.6
	RICE	0.0	0.0	-7.6	-7.6	-7.5	-7.4
	OILS	0.0	0.0	1.9	2.0	2.9	3.8

Table B.6: - continued -

		BA	SC1	SC2	SC3	SC4	SC5
BUR	SUGA	0.0	0.0	-23.5	-26.9	11.3	35.6
	MILK	0.0	0.0	29.3	27.7	12.5	-4.9
	BEEF	0.0	0.0	229.6	224.6	143.5	53.8
	PORK	0.0	0.0	51.7	51.7	40.0	29.4
	POUL	0.0	0.0	-13.3	-13.5	-30.2	-43.9
	COTT	0.0	0.2	31.9	31.6	27.6	22.6
	TOBA	0.0	0.5	-16.0	-15.7	-13.7	-10.4
	OLIO	0.0	1.7	-10.5	-6.8	-8.6	-6.3
	APPL	0.0	0.0	-19.2	-19.3	-21.6	-22.8
	ORAN	0.0	0.0	3.3	2.5	-5.7	-13.1
	TOMA	0.0	0.0	-45.4	-44.0	-44.6	-43.7
MOR	WHEA	0.0	0.0	0.0	-25.6	-14.8	-27.8
	COAR	0.0	0.0	-0.3	-3.7	-3.1	-7.7
	RICE	0.0	0.0	0.0	-37.9	-29.0	-56.6
	OILS	0.0	0.0	0.0	-0.7	0.3	0.6
	SUGA	0.0	0.0	0.0	85.0	4.1	10.1
	MILK	0.0	0.0	-0.1	-30.2	-30.0	-58.3
	BEEF	0.0	0.0	-0.4	220.1	7.1	22.0
	PORK	0.0	0.0	-0.5	29.7	6.0	13.4
	POUL	0.0	0.0	0.3	-45.5	-38.6	-68.2
	COTT	0.0	0.3	0.2	4.9	4.5	10.4
	TOBA	0.0	0.7	0.9	-15.2	-3.6	-7.2
	OLIO	0.0	2.0	2.0	-39.0	-18.6	-38.7
	APPL	0.0	0.0	0.1	-48.8	-25.7	-48.9
	ORAN	0.0	0.0	0.0	26.4	7.1	15.4
	TOMA	0.0	0.0	0.5	7.0	4.3	8.2
TUR	WHEA	0.0	0.0	0.0	12.3	2.0	4.4
	COAR	0.0	0.0	-0.3	-19.9	-11.5	-21.8
	RICE	0.0	0.0	0.0	46.9	3.4	9.4
	OILS	0.0	0.0	0.0	-22.8	-10.9	-21.7
	SUGA	0.0	0.0	0.1	41.3	-6.4	-11.7
	MILK	0.0	0.0	-0.6	70.0	-0.6	2.7
	BEEF	0.0	-0.1	-1.2	-19.7	-28.5	-59.7
	PORK	0.0	0.0	-0.2	22.3	2.8	6.2
	POUL	0.0	0.0	0.2	39.0	-9.0	-14.7
	COTT	0.0	0.4	0.4	-0.4	2.3	5.7
	TOBA	0.0	1.1	1.4	-20.2	-3.9	-8.3
	OLIO	0.0	2.7	2.7	-22.2	-9.5	-21.7
	APPL	0.0	0.0	0.1	12.0	5.4	11.8
	ORAN	0.0	0.0	0.0	26.3	7.3	15.7

Annex

Table B.6: - continued -

		BA	SC1	SC2	SC3	SC4	SC5
TUR	TOMA	0.0	0.0	0.7	11.1	6.7	13.0
MPC	WHEA	0.0	0.0	0.0	-6.3	-6.2	-11.9
	COAR	0.0	0.0	-0.4	5.8	0.0	0.8
	RICE	0.0	0.0	0.0	48.1	1.9	5.2
	OILS	0.0	0.0	0.0	-0.2	0.3	0.6
	SUGA	0.0	0.0	0.1	80.0	11.8	29.5
	MILK	0.0	0.0	-0.2	41.6	-12.5	-23.7
	BEEF	0.0	0.0	-0.4	190.5	1.2	8.1
	PORK	0.0	0.0	-0.4	26.0	4.5	10.2
	POUL	0.0	0.0	0.1	35.8	-10.4	-18.7
	COTT	0.0	0.3	0.3	0.8	2.1	5.0
	TOBA	0.0	0.6	0.8	-2.8	2.3	5.1
	OLIO	0.0	2.2	2.3	-33.6	-15.7	-33.2
	APPL	0.0	0.0	0.1	-0.1	-1.6	-2.9
	ORAN	0.0	0.0	0.0	12.3	1.7	3.9
	TOMA	0.0	0.0	0.5	-54.3	-29.4	-53.7

Source: Own simulations with AGRISIM

Table B.7: Changes in production shares relative to the world supply (in %)

		BA	SC1	SC2	SC3	SC4	SC5
Mediterra-nean EU Member States	WHEA	2.3	2.3	2.3	2.3	2.3	2.3
	COAR	2.6	2.7	2.7	2.7	2.6	2.6
	RICE	0.4	0.4	0.4	0.4	0.4	0.4
	OILS	1.3	1.3	1.3	1.3	1.3	1.3
	SUGA	2.1	2.1	2.1	2.1	2.1	2.0
	MILK	3.6	3.6	3.6	3.6	3.6	3.6
	BEEF	2.8	2.9	2.9	2.9	2.6	2.2
	PORK	5.2	5.2	5.1	5.1	4.9	4.7
	POUL	3.3	3.3	3.3	3.3	2.9	2.5
	COTT	2.7	2.6	2.6	2.6	2.5	2.5
	TOBA	5.0	4.8	4.8	4.8	4.8	4.8
	OLIO	77.6	77.3	77.3	78.7	78.0	78.7
	APPL	6.1	6.1	6.1	6.1	6.1	6.0
	ORAN	9.2	9.2	9.2	9.1	8.9	8.6
	TOMA	11.4	11.4	11.4	11.5	11.5	11.6
E12	WHEA	12.3	12.3	12.3	12.3	12.3	12.4
	COAR	7.3	7.3	7.3	7.3	7.2	7.2
	RICE	0.0	0.0	0.0	0.0	0.0	0.0
	OILS	7.9	7.9	7.9	7.9	7.9	7.9

Table B.7: - continued -

		BA	SC1	SC2	SC3	SC4	SC5
E12	SUGA	9.8	9.8	9.8	9.8	9.7	9.5
	MILK	17.9	17.9	17.9	17.9	17.8	17.6
	BEEF	8.8	8.8	8.8	8.8	8.2	7.0
	PORK	14.8	14.8	14.8	14.8	14.2	13.7
	POUL	9.8	9.8	9.8	9.7	8.6	7.5
	COTT	0.0	0.0	0.0	0.0	0.0	0.0
	TOBA	0.7	0.7	0.7	0.7	0.7	0.7
	OLIO	1.2	1.2	1.2	1.2	1.2	1.2
	APPL	10.5	10.5	10.5	10.5	10.4	10.3
	ORAN	0.4	0.4	0.4	0.4	0.3	0.3
	TOMA	2.8	2.8	2.8	2.8	2.8	2.8
MOR	WHEA	0.6	0.6	0.6	0.5	0.6	0.6
	COAR	0.1	0.1	0.1	0.1	0.1	0.1
	RICE	0.0	0.0	0.0	0.0	0.0	0.0
	OILS	0.0	0.0	0.0	0.0	0.0	0.0
	SUGA	0.4	0.4	0.4	0.4	0.4	0.4
	MILK	0.2	0.2	0.2	0.2	0.2	0.2
	BEEF	0.3	0.3	0.3	0.3	0.3	0.3
	PORK	0.0	0.0	0.0	0.0	0.0	0.0
	POUL	0.4	0.4	0.4	0.3	0.3	0.2
	COTT	0.0	0.0	0.0	0.0	0.0	0.0
	TOBA	0.1	0.1	0.1	0.1	0.1	0.1
	OLIO	1.9	2.0	2.0	1.8	1.9	1.8
	APPL	0.4	0.4	0.4	0.3	0.3	0.3
	ORAN	1.1	1.1	1.1	1.3	1.2	1.2
	TOMA	0.8	0.8	0.8	0.9	0.9	0.9
TUR	WHEA	3.3	3.3	3.3	3.3	3.3	3.3
	COAR	1.1	1.1	1.1	1.0	1.1	1.1
	RICE	0.1	0.1	0.1	0.1	0.1	0.1
	OILS	0.5	0.5	0.5	0.5	0.5	0.5
	SUGA	1.7	1.7	1.7	1.7	1.7	1.6
	MILK	1.6	1.6	1.6	2.0	1.6	1.6
	BEEF	0.6	0.6	0.6	0.5	0.5	0.5
	PORK	0.0	0.0	0.0	0.0	0.0	0.0
	POUL	0.9	0.9	0.9	1.0	0.9	0.8
	COTT	4.3	4.3	4.3	4.2	4.3	4.3
	TOBA	2.3	2.3	2.3	2.3	2.3	2.3
	OLIO	3.3	3.4	3.4	3.2	3.3	3.3
	APPL	4.2	4.2	4.2	4.4	4.3	4.4
	ORAN	2.0	2.0	2.0	2.2	2.1	2.1

Annex

Table B.7: - continued -

		BA	SC1	SC2	SC3	SC4	SC5
TUR	TOMA	8.0	8.0	8.0	8.4	8.3	8.5
MPC	WHEA	2.5	2.5	2.5	2.4	2.5	2.5
	COAR	1.6	1.6	1.6	1.6	1.7	1.7
	RICE	0.9	0.9	0.9	0.9	0.9	0.9
	OILS	0.1	0.1	0.1	0.1	0.1	0.1
	SUGA	1.2	1.2	1.2	1.2	1.2	1.2
	MILK	1.7	1.7	1.7	1.9	1.6	1.5
	BEEF	1.5	1.5	1.5	1.8	1.5	1.5
	PORK	0.0	0.0	0.0	0.0	0.0	0.0
	POUL	2.4	2.4	2.4	2.6	2.3	2.2
	COTT	3.3	3.3	3.3	3.3	3.3	3.3
	TOBA	0.9	0.9	0.9	0.9	0.9	0.9
	OLIO	12.2	12.4	12.4	11.3	11.9	11.3
	APPL	2.1	2.1	2.1	2.1	2.1	2.1
	ORAN	4.9	4.9	4.9	5.1	4.9	5.0
	TOMA	9.3	9.3	9.4	5.6	7.5	5.7

Source: Own simulations with AGRISIM

Table B.8: Change in production shares of the Mediterranean Member States relative to the EU-27 supply (in %)

		BA	SC1	SC2	SC3	SC4	SC5
ESP	WHEA	3.9	3.9	3.9	3.9	3.9	4.0
	COAR	7.8	7.9	7.8	7.8	7.9	7.9
	RICE	34.2	34.2	34.2	34.2	34.2	34.2
	OILS	5.0	5.1	5.1	5.1	5.1	5.2
	SUGA	5.6	5.6	5.5	5.6	5.6	5.6
	MILK	4.6	4.6	4.6	4.6	4.6	4.6
	BEEF	7.5	7.6	7.5	7.5	7.5	7.3
	PORK	13.7	13.7	13.7	13.7	13.5	13.2
	POUL	9.4	9.4	9.4	9.4	9.3	9.2
	COTT	18.2	18.1	18.1	18.1	18.1	18.1
	TOBA	9.6	9.9	9.9	9.9	9.9	9.9
	OLIO	57.7	57.9	57.9	57.9	57.9	57.9
	APPL	6.9	6.9	7.0	7.0	6.9	6.8
	ORAN	47.3	47.3	47.3	47.4	47.2	47.1
	TOMA	22.3	22.3	22.8	22.9	22.9	23.0
GRE	WHEA	2.1	2.1	2.1	2.1	2.1	2.2
	COAR	1.7	1.7	1.6	1.6	1.6	1.6
	RICE	5.8	5.8	5.8	5.8	5.8	5.8

Table B.8: - continued -

		BA	SC1	SC2	SC3	SC4	SC5
GRE	OILS	0.1	0.1	0.1	0.1	0.1	0.1
	SUGA	1.8	1.8	1.8	1.8	1.8	1.9
	MILK	1.2	1.2	1.2	1.2	1.2	1.2
	BEEF	0.7	0.7	0.7	0.7	0.7	0.6
	PORK	0.6	0.6	0.6	0.6	0.6	0.6
	POUL	1.3	1.3	1.3	1.3	1.2	1.2
	COTT	81.3	81.4	81.3	81.3	81.3	81.3
	TOBA	31.1	31.0	31.1	31.1	31.1	31.2
	OLIO	19.5	19.2	19.2	19.3	19.3	19.3
	APPL	1.8	1.8	1.8	1.8	1.8	1.9
	ORAN	17.2	17.2	17.2	17.1	17.3	17.5
	TOMA	10.9	10.9	11.1	11.2	11.2	11.2
ITA	WHEA	5.1	5.2	5.3	5.3	5.3	5.3
	COAR	8.0	8.1	8.0	8.0	7.9	7.8
	RICE	49.6	49.6	49.6	49.6	49.6	49.6
	OILS	4.7	4.7	4.7	4.7	4.7	4.7
	SUGA	7.5	7.5	7.5	7.5	7.5	7.6
	MILK	8.0	8.0	8.0	8.0	8.0	8.0
	BEEF	12.9	13.0	12.9	12.9	12.8	12.4
	PORK	6.9	6.9	6.9	6.9	6.9	6.9
	POUL	10.0	10.0	10.0	10.0	9.9	9.7
	COTT	0.0	0.0	0.0	0.0	0.0	0.0
	TOBA	29.8	28.8	28.8	28.8	29.0	29.1
	OLIO	21.2	21.2	21.2	21.2	21.2	21.2
	APPL	16.9	16.9	16.9	16.9	16.8	16.6
	ORAN	31.2	31.2	31.2	31.2	31.2	31.2
	TOMA	39.1	39.1	39.9	39.7	39.8	39.7

Source: Own simulations with AGRISIM

Table B.9: Change in welfare (deviations from BA in US$ million)

		SC1	SC2	SC3	SC4	SC5
ESP	producer surplus	-35	-41	93	-490	-991
	quota owner surplus	25	18	-7	-318	-710
	consumer surplus	-21	-7	-39	885	2044
	budget (taxpayers effect)[1]	51	54	57	267	175
	total[1]	20	24	105	344	518
	budget (taxpayers effect)[2]	-12	-12	-5	12	6
	total[2]	-44	-42	43	88	349

Annex

Table B.9: - continued -

		SC1	SC2	SC3	SC4	SC5
GRE	producer surplus	-175	-174	-118	-192	-202
	quota owner surplus	27	25	18	-52	-139
	consumer surplus	-11	-9	-36	210	519
	budget (taxpayers effect)[1]	182	183	184	95	-117
	total[1]	22	25	48	61	61
	budget (taxpayers effect)[2]	-174	-174	-173	-37	155
	total[2]	-333	-331	-310	-71	333
ITA	producer surplus	-145	-143	-44	-519	-969
	quota owner surplus	9	-4	-45	-601	-1328
	consumer surplus	-27	-9	-52	1330	3182
	budget (taxpayers effect)[1]	151	153	160	54	-593
	total[1]	-13	-2	19	263	292
	budget (taxpayers effect)[2]	-87	-84	-76	405	890
	total[2]	-251	-239	-217	615	1775
E12	producer surplus	-35	-70	-32	-3213	-6293
	quota owner surplus	-16	-125	-481	-4957	-10508
	consumer surplus	-18	107	356	8662	19292
	budget (taxpayers effect)[1]	52	65	108	1592	825
	total[1]	-17	-24	-49	2084	3315
	budget (taxpayers effect)[2]	259	275	309	668	640
	total[2]	190	186	153	1161	3130
E10	producer surplus	0	-23	-47	-1143	-2177
	quota owner surplus	-1	-23	-93	-764	-1538
	consumer surplus	-2	24	69	1290	2773
	budget (taxpayers effect)[1]	0	21	74	1153	1759
	total[1]	-2	0	4	537	818
	budget (taxpayers effect)[2]	14	-5	-55	-1048	-1691
	total[2]	11	-26	-125	-1664	-2632
EU-25	producer surplus	-391	-451	-147	-5557	-10633
	quota owner surplus	45	-109	-608	-6692	-14223
	consumer surplus	-79	107	299	12377	27811
	budget (taxpayers effect)	435	475	583	3161	2048
	total	10	22	127	3290	5004

Table B.9: - continued -

		SC1	SC2	SC3	SC4	SC5
BUR	producer surplus	0	2527	2527	2135	1773
	quota owner surplus	0	587	567	421	245
	consumer surplus	-1	-160	-154	240	704
	budget (taxpayers effect)	0	-3150	-3135	-2840	-2708
	total	-1	-196	-195	-43	14
MOR	producer surplus	3	2	125	-455	-857
	quota owner surplus	0	0	0	0	0
	consumer surplus	-3	-2	27	522	1026
	budget (taxpayers effect)	0	-1	-226	-36	-163
	total	0	0	-75	31	7
TUR	producer surplus	8	0	1332	-227	-386
	quota owner surplus	0	0	0	0	0
	consumer surplus	-8	0	-1002	283	552
	budget (taxpayers effect)	-2	1	-391	61	39
	total	-2	2	-61	117	204
MPC	producer surplus	20	17	5012	-1458	-2608
	quota owner surplus	0	0	0	0	0
	consumer surplus	-21	-14	-5490	1594	3010
	budget (taxpayers effect)	-4	-8	-341	54	-338
	total	-6	-4	-819	191	65

Notes: [1] Without considering intra-community financial flows
[2] Considering intra-community financial flows
Source: Own simulations with AGRISIM

For the quotation used throughout the tables in the Annex the reader should refer to Table 5.1.

Schriften zur Internationalen Entwicklungs- und Umweltforschung

Herausgegeben vom
Zentrum für internationale
Entwicklungs- und
Umweltforschung
der Justus-Liebig-Universität Gießen

Band 1 Hans-Rimbert Hemmer / Rainer Wilhelm: Fighting Poverty in Developing Countries. Principles for Economic Policy. 2000.

Band 2 Lorenz King / Martin Metzler / Tong Jiang (eds.): Flood Risks and Land Use Conflicts in the Yangtze Catchment, China and at the Rhine River, Germany. 2001.

Band 3 Ingrid-Ute Leonhäuser (ed.): Women in the Context of International Development and Cooperation. Review and Perspectives. Selected Papers and Abstracts presented at the Justus-Liebig-University Gießen 26.–28. October 2000. 2002

Band 4 Margit Schratzenstaller: Internationale Mobilität von und internationaler fiskalischer Wettbewerb um Direktinvestitionen. 2002.

Band 5 Armin Bohnet u.a.: Theoretische Grundlagen und praktische Gestaltungsmöglichkeiten eines Finanzausgleichssystems für die VR China. Unter Mitwirkung von Chen Biyan, Chen Shixin, Ge Licheng, Ge Naixu, Ge Zhuying, Ma Shuanyou, Markus Peplau, Yang Zhigang, Zhu Qiuxia. 2003.

Band 6 Armin Bohnet / Matthias Höher (eds.): The Role of Minorities in the Development Process. 2004.

Band 7 Thi Phuong Hoa Nguyen: Foreign Direct Investment and its Contributions to Economic Growth and Poverty Reduction in Vietnam (1986–2001). 2004.

Band 8 Andreas Böcker / Roland Herrmann / Michael Gast / Jana Seidemann: Qualität von Nahrungsmitteln. Grundkonzepte, Kriterien, Handlungsmöglichkeiten. 2004.

Band 9 Christina Mönnich: Tariff Rate Quotas and Their Administration. Theory, Practice and an Econometric Model for the EU. 2004.

Band 10 Reimund Seidelmann / Ernst Giese (eds.): Cooperation and Conflict Management in Central Asia. 2004.

Band 11 Claudia Ohly: Das Steuersystem im ungarischen Transformationsprozess. Ein Beitrag zur Transformationstheorie. 2004.

Band 12 Nicole Mau: Umweltzertifikate. Der Einsatz von Umweltzertifikaten in der Landwirtschaft am Beispiel klimarelevanter Gase. 2005.

Band 13 P. Michael Schmitz (Hrsg.): Water and Sustainable Development. 2005.

Band 14 Ira Pawlowski: Die Wettbewerbsfähigkeit der ukrainischen Milchwirtschaft. Messung von Marktverzerrung und Politikeinfluß im Transformationsprozeß. 2005.

Band 15 Kirsten Westphal (ed.): A Focus on EU-Russian Relations. Towards a close partnership on defined road maps? 2005.

Band 16 Andreas Langenohl / Kirsten Westphal (eds.): Conflicts in a Transnational World. Lessons from Nations and States in Transformation. 2006.

Band 17 Rosemarie von Schweitzer: Home Economics Science and Arts. Managing Sustainable Everyday Life. 2006.

Band 18 Dörthe List: Regionale Kooperation in Zentralasien. Hindernisse und Möglichkeiten. 2006.

Band 19 Michael Gast: Determinanten ausländischer Direktinvestitionen. OECD-Länder als Investoren und besonderer Aspekte der Ernährungswissenschaft. 2007.

Band 20 Kim Schmitz: Die Bewertung von Multifunktionalität der Landschaft mit diskreten Choice Experimenten. 2008.

Band 21 Kerstin Kötschau / Thilo Marauhn (eds.): Good Governance and Developing Countries. Interdisciplinary Perspectives. 2008.

Band 22 Armin Bohnet (ed.): Poland on its Way to a Federal State? 2008.

Band 23 Johannes Harsche: Regionale Inzidenz und ökonomische Bestimmungsgrößen der Gemeinsamen Europäischen Agrarpolitik. 2009.

Band 24 Aikaterini Kavallari: Agricultural Trade Policy Reforms and Trade Liberalisation in the Mediterranean Basin. A Partial Equilibrium Analysis of Regional Effects on the EU-27 and on the Mediterranean Partner Countries. 2009.

Band 25 Marc Christopher Kramb: Sanitäre und phytosanitäre Handelsbeschränkungen unter dem Einfluss des WTO-Abkommens. Ein Gravitationsansatz unter besonderer Berücksichtigung des EU-Rindfleischsektors. 2009.

www.peterlang.de

www.ingramcontent.com/pod-product-compliance
Ingram Content Group UK Ltd.
Pitfield, Milton Keynes, MK11 3LW, UK
UKHW021829210426
5322IPUK00004B/103